Synthesis Lectures on Renewable Energy Technologies

The series, Synthesis Lectures on Renewable Energy Technologies publishes concise books, focused on technologies that harness energy from naturally occurring sources, such as sunlight, wind, water, geothermal heat, and biofuels from organic materials. These renewable energy technologies play a crucial role in transitioning away from fossil fuels, helping to mitigate the effects of climate change, and promoting a sustainable energy supply.

P. Arjun Suresh · K. V. Arun Kumar ·
Ann Rose Abraham · A. K. Haghi

Perovskite Solar Cells Technology

Next Generation Clean Energy Solution

P. Arjun Suresh
Department of Physics
CMS College (Autonomous)
Kottayam, Kerala, India

Ann Rose Abraham
Department of Physics
Sacred Heart College (Autonomous)
Thevara, Kerala, India

K. V. Arun Kumar
Department of Physics
CMS College (Autonomous)
Kottayam, Kerala, India

A. K. Haghi
Edinburgh, UK

Department of Chemistry
Institute of Molecular Sciences
University of Coimbra
Coimbra, Portugal

ISSN 2690-5000   ISSN 2690-5019 (electronic)
Synthesis Lectures on Renewable Energy Technologies
ISBN 978-3-031-90749-4   ISBN 978-3-031-90750-0 (eBook)
https://doi.org/10.1007/978-3-031-90750-0

© The Editor(s) (if applicable) and The Author(s), under exclusive license to Springer
Nature Switzerland AG 2026

This work is subject to copyright. All rights are solely and exclusively licensed by the Publisher, whether the whole or part of the material is concerned, specifically the rights of translation, reprinting, reuse of illustrations, recitation, broadcasting, reproduction on microfilms or in any other physical way, and transmission or information storage and retrieval, electronic adaptation, computer software, or by similar or dissimilar methodology now known or hereafter developed.
The use of general descriptive names, registered names, trademarks, service marks, etc. in this publication does not imply, even in the absence of a specific statement, that such names are exempt from the relevant protective laws and regulations and therefore free for general use.
The publisher, the authors and the editors are safe to assume that the advice and information in this book are believed to be true and accurate at the date of publication. Neither the publisher nor the authors or the editors give a warranty, expressed or implied, with respect to the material contained herein or for any errors or omissions that may have been made. The publisher remains neutral with regard to jurisdictional claims in published maps and institutional affiliations.

This Springer imprint is published by the registered company Springer Nature Switzerland AG
The registered company address is: Gewerbestrasse 11, 6330 Cham, Switzerland

If disposing of this product, please recycle the paper.

Preface

Perovskite solar cells are an emerging field in photovoltaics. Unlike other solar cell technologies such as silicon solar cells, perovskite solar cells have shown a remarkable increase in efficiency within a short time. This rapid improvement highlights their importance. Despite their higher power conversion efficiency (PCE) values, commercialization remains challenging due to their instability and the difficulties associated with large-scale fabrication. This book provides a comprehensive summary of the current status and development of perovskite solar cells (PSCs). It is divided into six main chapters.

Chapter 1 offers a brief introduction to solar cells, covering the basic working principles, classifications, history, and evolution of PSCs. This chapter provides fundamental knowledge about solar cell technologies, enabling readers to understand the subsequent chapters.

Chapter 2 explores the structure and properties of perovskite materials, comparing perovskite solar cells with other photovoltaic technologies. It also provides a summary of various perovskite materials that can be used as active layers in PSCs.

Chapter 3 discusses the different layers and their roles in the efficient functioning of perovskite solar cells. This chapter delves into various techniques used for PSC fabrication and the principles underlying their operation.

Chapter 4 includes a detailed analysis of factors influencing efficiency. It examines different materials that can be used as hole transport layers (HTL), electron transport layers (ETL), and electrodes to improve the efficiency and stability of PSCs. Given graphene's popularity due to its conductivity and hydrophobic nature, this chapter also includes an in-depth analysis of how to use graphene in PSCs to enhance efficiency and stability.

Chapter 5 focuses on the challenges in preparing PSCs and the major techniques that can be employed for their commercialization. Stability is one of the major factors hindering the commercialization of PSCs. Encapsulation is a key technique that can be used to enhance stability. This chapter explains various encapsulation methods for stability improvement. Since perovskite solar cells use lead for fabricating their active layer, which poses environmental hazards, this chapter also discusses the toxicity of lead in detail and explores ways to reduce or replace its use to make PSCs more environmentally friendly.

Chapter 6 is dedicated to the future of PSCs. It discusses ways to make PSCs more efficient and eco-friendlier, the future of tandem solar cell technologies, the potential for flexible PSCs, and a detailed study of inverted perovskite solar cells.

As the global community strives to transition to cleaner energy sources, the significance of perovskite solar cells cannot be overstated. We hope that this book will not only provide valuable insights but also inspire readers to contribute to the development and deployment of this transformative technology.

Kottayam, India	P. Arjun Suresh
Kottayam, India	K. V. Arun Kumar
Thevara, India	Ann Rose Abraham
Edinburgh, UK/Coimbra, Portugal	A. K. Haghi

Contents

1 Introduction to Solar Energy 1
   1.1 Importance of Renewable Energy 1
   1.2 Brief Study About Solar Energy 2
   1.3 Classification of Solar Cells 2
      1.3.1 First-Generation Solar Cells 2
      1.3.2 Second-Generation Solar Cells 3
      1.3.3 Third-Generation Solar Cells 4
   1.4 Basic Working of Solar Cells 6
   1.5 History and Evolution of Perovskite Solar Cells 8
   1.6 Drawbacks of Perovskite Solar Cells 10
   References 11

2 Perovskite Materials 13
   2.1 Structure of Perovskites Material 13
   2.2 Major Perovskite Material 14
      2.2.1 $MA_xFA_{1-x}PbI_3$ Based System 15
      2.2.2 $Cs_xFA_{1-x}PbI_3$ System 16
      2.2.3 $Rb_xFA_{1-x}PbI_3$ System 17
      2.2.4 $MAPbI_{3-x}Br_x$ System 17
   2.3 Comparison with Traditional Solar Cell 18
      2.3.1 Crystalline Silicon Solar Cell (c-Si) 18
      2.3.2 Gallium Arsenide (GaAs) 19
      2.3.3 Multi-junction Solar Cells (III–V) 19
      2.3.4 Perovskite Solar Cell 20
   References 21

3 Fabrication of Perovskite Solar Cell 23
   3.1 Device Structure 23
      3.1.1 Electron Transport Layer 24
      3.1.2 Hole Transport Layer (HTL) 25

	3.2	Solution Processing Methods	26
		3.2.1 Spin-Coating Method	26
		3.2.2 Drop-Casting Method	27
		3.2.3 Roll to Roll Printing	27
		3.2.4 Slot Die Coating	28
		3.2.5 Spray Coating	29
		3.2.6 Blade Coating Method	30
	3.3	Vapor-Based Techniques	31
		3.3.1 Vapor-Assisted Deposition	31
		3.3.2 Chemical Vapor Deposition	32
		3.3.3 Physical Vapor Deposition	32
		3.3.4 Vapor-Assisted Solution Process	33
		References	34
4	**Efficiency Enhancement of PSC**	35
	4.1	Tolerance Factor	35
	4.2	Current Efficiency and Scope of Enhancement	36
	4.3	HTL and ETL Dependence of Efficiency	37
		4.3.1 Selection of HTL	38
	4.4	Graphene as Conducting Electrode, HTL and ETL	44
		4.4.1 Conducting Electrodes	44
		4.4.2 Materials Employed in Hole Transport	46
		4.4.3 Material for Transporting Electrons	47
	4.5	Influence of Graphene in Perovskite Materials	49
		References	50
5	**Challenges of Preparing PSC**	51
	5.1	Difficulties in Preparing PSC	51
	5.2	Large Area Fabrication of Perovskite	53
		5.2.1 Soft Cover Deposition	54
		5.2.2 Inkjet Printing	54
		5.2.3 Doctor-Blade Coating	55
	5.3	Dependence of Humidity and Light in the Fabrication of Perovskite	56
	5.4	Stability of PSC	58
		5.4.1 Effect of Encapsulation in PSC	59
	5.5	Addressing Toxicity of Lead-Based PSC	61
		References	63

6 Future of PSC ... 65
6.1 Tandem Solar Cells ... 65
6.1.1 Tandem Cell Configurations ... 66
6.2 Pros and Cons of Perovskite Material in Tandem Solar Cells ... 69
6.3 Efficiency Progress of Tandem Solar Cells ... 71
6.4 Flexible Solar Cells ... 73
6.4.1 Flexible Substrates ... 73
6.5 Inverted Perovskite Solar Cells ... 74
6.5.1 Different Configurations of IPSC ... 76
6.6 Lead Free PSC ... 79
References ... 80

About the Authors

P. Arjun Suresh is a Ph.D. research scholar at CMS College (autonomous), Kottayam affiliated to MG University, Kottayam, Kerala, specializing in perovskite solar cells. He holds an M.Phil. in Physics from Sacred Heart College, Thevara, where he completed his project work at IIT Madras. Arjun has a deep knowledge of solid-state physics, materials science, and nanotechnology. His research interests encompass the electrocaloric effect, graphene, and perovskite solar cells. He has authored multiple high-impact publications and book chapters in leading Scopus-indexed journals and books. His recent work, published in the Journal of Materials Science: Materials in Electronics (Springer, 2024), introduces innovative fabrication methods for the $FAPbI_3$ active layer in solar cells, advancing renewable energy technologies. Dedicated to enhancing solar cell efficiency and performance, Arjun continues to explore novel materials and fabrication techniques. Through his research and publications, he strives to make meaningful contributions to sustainable energy solutions and the broader scientific community.

K. V. Arun Kumar, Ph.D. is working as an assistant professor in the Department of Physics at CMS College (Autonomous), Kottayam, Kerala, India, since 2015. He has 17 years of research experience and 10 years of teaching experience. He has 23 international journal publications and 8 book chapters. He has worked at prestigious national institutes, including the Center for Materials for Electronics Technology (C-MET), Thrissur (2006–2007), School of Pure and Applied Physics, Mahatma Gandhi University, Kottayam (2007–2013), and CSIR-National Institute for Interdisciplinary Science and Technology (NIIST), Thiruvananthapuram (2013–2014). His research interests primarily focus on sol-gel synthesized metal/rare earth-doped glasses. His work includes the development of sol-gel SiO_2, SiO_2-TiO_2, and SiO_2-ZrO_2 multi-component glasses, piezoelectric actuators, sol-gel $LaPO_4$-based ceramic nanocomposite materials, DSSC solar cells, perovskite materials for solar cells applications, plasmonic metal nanoparticles, rare earth-doped fluorescence studies, dielectric studies, and colorimetric studies.

Ann Rose Abraham, Ph.D. *Assistant Professor, Sacred Heart College (Autonomous), Thevara, Kochi, India.*

Dr. Ann Rose Abraham is currently an assistant professor of Physics at Sacred Heart College (Autonomous), Thevara, Cochin, India. Her academic journey has been marked by notable achievements and conducting ongoing research at the Materials Research Laboratory (MRL), Sacred Heart College, Kochi. Dr. Abraham received her M.Sc., M.Phil., and Ph.D. degrees in Physics from the School of Pure and Applied Physics, Mahatma Gandhi University, Kerala, India. She has authored, co-authored, edited, or co-edited more than 60 publications, including books, book chapters, and papers in peer-reviewed journals. She is a reviewer of many international journals. She has several publications to her credit in many peer-reviewed high impact journals of international repute, such as Elsevier, Taylor and Francis, Springer, *Journal of Physical Chemistry C*, *Physical Chemistry Chemical Physics*, *New Journal of Chemistry*, *Philosophical Magazine*, etc. She has research experience at various national institutes including Bose Institute, SAHA Institute of Nuclear Physics, UGC-DAE CSR Centre, Kolkata, and has collaborations with various international laboratories, such as the Université de Lorraine (France), University of Johannesburg, Institute of Physics Belgrade, etc. She is the recipient of young researcher award in physics, a prestigious forum to showcase intellectual capability. She has expertise in the field of materials science, nanomagnetic materials, multiferroics, polymeric nanocomposites, biomaterials, etc.

A. K. Haghi, Ph.D. Research Associate, Department of Chemistry, Institute of Molecular Sciences, University of Coimbra, Coimbra, Portugal.

A. K. Haghi is a retired professor and has written, co-written, edited, or co-edited more than 1000 publications, including books, book chapters, and papers in refereed journals with over 4200 citations and h-index of 35, according to the Google Scholar database. Professor Haghi holds a B.Sc. in urban and environmental engineering from the University of North Carolina (USA) and holds two M.Sc. degrees, one in mechanical engineering from North Carolina State University (USA) and another one in applied mechanics, acoustics, and materials from the Université de Technologie de Compiègne (France). He was awarded a Ph.D. in engineering sciences at Université de Franche-Comté (France).

Professor Haghi's extensive educational background and supervisory roles underscore his expertise and contributions to the field of engineering sciences. He is appointed as Honorary Research Associate (HRA) at University of Coimbra, Portugal. He is a regular reviewer of leading international journals.

List of Figures

Fig. 1.1	Basic structure of p-n junction solar cells	7
Fig. 1.2	I-V and P-V charecteristics of solar cell (open access)	8
Fig. 1.3	Efficiency versus year graph, growth of perovskite solar cells (Copyright)	9
Fig. 2.1	Perovskite structure (open access)	14
Fig. 2.2	Perovskite structured hybrid halide (copyright)	15
Fig. 3.1	Layers of perovskite solar cells	24
Fig. 3.2	Classification of PSC fabrication technique	26
Fig. 3.3	Steps in the spin-coating method	27
Fig. 3.4	Schematic diagram of drop-casting method	28
Fig. 3.5	Classification of Roll-to-Roll compatible fabrication techniques	28
Fig. 3.6	Schematic representation of slot die coating	29
Fig. 3.7	Illustration of spray-coating method	30
Fig. 3.8	Schematic diagram of blade coating method	30
Fig. 3.9	Schematic diagram of vapor-assisted deposition	31
Fig. 3.10	Perovskite film fabricated using chemical vapor deposition	32
Fig. 3.11	Schematic diagram of physical vapor deposition	33
Fig. 3.12	Illustrates vapor assisted solution process (open access)	34
Fig. 4.1	Perovskite structure (Open access)	36
Fig. 4.2	Energy level diagram of inorganic HTLs (Copyright)	38
Fig. 4.3	Structure of graphene	44
Fig. 5.1	Illustration of one step method	52
Fig. 5.2	Represents the growth mechanisms occurs in two step deposition process **a** interfacial reaction **b** dissolution and recrystallization (Open access)	53
Fig. 5.3	Illustration of the SCD technique	54
Fig. 5.4	Illustration of inkjet printing	55
Fig. 5.5	Illustration of various steps in doctor blade method (Copyright)	56

Fig. 5.6	Illustrates the various encapsulation methods **a** thin film encapsulation (TFE) **b** edge seal method (copyright)	60
Fig. 6.1	Represents the schematic diagram of **a** four terminal tandem device **b** two terminal tandem device (Open access)	66
Fig. 6.2	Classifications of tandem solar cells	66
Fig. 6.3	Tandem configurations of four terminal tandem cell	67
Fig. 6.4	Tandem configurations of two terminal tandem device	67
Fig. 6.5	Tandem configurations of reflective tandem cells	68
Fig. 6.6	Tandem configurations of series–parallel tandem cells	69
Fig. 6.7	Illustrates a schematic diagram four-terminal tandem device	72
Fig. 6.8	Schematic diagram of 2T perovskite/Si solar cells (Open access)	72
Fig. 6.9	Flexible solar cell with polymer substrate (Copyright)	75
Fig. 6.10	Illustrates the schematic diagram of IPSCs	75
Fig. 6.11	Energy levels of different components IPSC (Copyright)	78

Introduction to Solar Energy

1.1 Importance of Renewable Energy

Nowadays humans are greatly dependent on electricity. And it becomes an unavoidable form of energy in our daily life. In the present scenario, nearly all the electric energy is yielded by from burning fossil fuels, and nearly 79.5% of energy is produced from non-renewable resources [1]. Renewable energy is becoming more and more significant, its clean and sustainable character has encouraged individuals to seriously consider it [2]. As we all know, fossil fuels are one of the non-replenishable sources of energy and it is depleting year by year. Also, it will cause serious environmental problems which we are facing right now. Fossil fuels have a high percentage of carbon contents and burning these will produce high amounts of carbon dioxide and many toxic gases which will lead to global warming [3]. A huge amount of energy is wasted when fossil fuels are burned to produce electricity [1]. The abundant and infinite supply of renewable energy sources makes it more promisable. They are green energy sources and have a much lower environmental impact than typical fossil fuel technology. Instead of being spent on costly energy imports in non-renewable energy resources, a major part of of investments in renewable energy is focused towards materials and labour for building and maintaining facilities [2]. Another significant benefit of renewable energy is its low carbon emissions, which significantly help in combating climate change caused by fossil fuel combustion. Since the cost of renewable energy is mostly dependent on the initial prices, it can stabilise the power prices up to some extent [2].

1.2 Brief Study About Solar Energy

As a solution to these environmental problems photovoltaic technologies were came, it is here for more than a century now. This technology is considered as one of the cleanest and safest methods for producing electricity. Since it doesn't produce any harmful gases this method should be considered. Unlike other methods, in photovoltaic method, the sunlight is being directly converted into electric energy [2]. Silicon solar cells are already ruling the market since it is efficient and it has a good lifetime of nearly 25 years. Considering other renewable energy technologies like wind energy, tidal energy etc. solar energy has a great lead, since it is less influenced by environmental factors like location, and temperature. Studies reveal that Earth can get 1575–49,387 EJ (exajoules) of radiation energy from the sun annually [1]. The silicon solar cell carries a huge disadvantage as it can't be easily manufactured or it is not economical. Researches are going on to find a better material having less cost of production, good efficiency and a high lifetime.

1.3 Classification of Solar Cells

There are three generations of solar cell technology.

1.3.1 First-Generation Solar Cells

The first-generation solar cells are crystalline Si wafer-based solar cells [4]. They are used as panels made of silicon solar cells [5]. They can be of two types

(i) Single crystalline/monocrystalline solar cells
(ii) Polycrystalline/multicrystalline solar cells.

Single-crystalline solar cells outperform multicrystalline ones in terms of efficiency [4, 6, 7]. They are the commercially available solar cells worldwide having the highest efficiency [8]. And these solar cells dominate 80% of the solar cell market [4].

The monocrystalline solar cells are manufactured by the process named as Czochralski process. This process involves cutting silicon crystal slices from large ingots, a procedure that requires detailed accuracy and results in high production costs [6]. Polycrystalline solar cells are fabricated by slicing out of a block of polycrystalline silicon which is formed by mixing several different crystals together [5]. These are also popular now. And the process of its making includes cooling mould filled with graphite. So manufacturing polycrystalline solar cells is more economical [6].

1.3.2 Second-Generation Solar Cells

The drawback concerning the first-generation solar cells is its higher production cost and difficulty in extracting pure silicon. To overcome this problem, thin film technology is adopted by depositing thin films of silicon (1 μm). The amount of silicon required is much less when compared to the wafer-based technology of first-generation solar cells [4]. The advantages of the second generation solar cells are (i) low material consumption, (ii) large area modules, (iii) low-temperature processes, (iv) tunable material properties, (v) transparent modules, and (vi) monolithic integration.

1.3.2.1 Amorphous Silicon Solar Cells (A-Si)
Amorphous Si solar cells are the most developed thin-film technologies among the others. The silicon used in these solar cells is in a non-crystalline form. While they are cost-effective, they are less efficient. A-Si solar cells are commonly used to power pocket calculators, buildings and remote facilities. Generally, the vapor deposition method is used to fabricate A-Si solar cells. A thin film of amorphous silicon, approximately one micrometre thick, is coated onto a glass or metal substrate. Deposition can also be performed at lower temperatures (around 75 °C), enabling coating on plastics. Flexibility is another advantage, allowing them to be incorporated into various structures. However, comparing to crystalline solar cells, its efficiency is lower. The second-generation solar cells have a crystal structure with a p-i-n layer configuration, whereas first-generation solar cells are based on p-n junction diodes [5, 9].

1.3.2.2 Cadmium Telluride Solar Cell (CdTe)
In cadmium telluride solar cells, the solar energy absorption and its conversion to electricity is done by a thin CdTe semiconductor layer. It is the sole thin film technology that is cheaper than the crystalline Si solar cells in multi-kilowatt systems. CdTe solar cells have advantages like lowest use of water, smallest carbon print, shortest energy payback duration against all other solar technologies. It is also considered as a very durable material and can fabricated using a wide range of methods. It has strong absorption coefficient and an optimum bandgap of 1.54 eV. Considering the bandgap it has low V_{OC} (open circuit voltage) of 0.85 eV, it should reach at least 1.2 V. The disadvantage is that the toxicity of cadmium (Cd), Cd being one of the most lethal and hazardous material. But CdTe is less harmful than elemental Cd, that too in case of acute exposure [9].

1.3.2.3 Copper Indium Gallium Selenide Solar Cells (CI(G)S)
Copper Indium Gallium Selenide Solar Cells (CIGS or CI(G)S or CIS cells) are formed by the deposition of a fine layer of copper (Cu), indium (In), gallium (Ga), and selenide (Se) onto a glass or plastic backing, which is then connected with electrodes on its back and front for current collection. The advantage of CIGS solar cells is that, due to its higher

absorption coefficient, they require only a much thinner film compared to other semiconductor materials. Also, the layers being so thin, can be deposited on flexible substrates to produce lightweight and highly flexible solar cells [5]. On flexible substrates based on ternary chalcopyrite Cu(In,Ga)Se$_2$, CIGS system achieves an efficiency of 20.8%, while 22.3% in the case of rigid substrates [4].

The main advantages of CIGS system are

- Thinner CIGS film allows to prepare flexible solar cells.
- Remarkable efficiency and low cost make CIGS devices more promisable.
- The high tolerance to radiation of the CIGS system makes it suitable for space applications.
- It can be synthesised using vacuum and non-vacuum deposition processes [9].

1.3.3 Third-Generation Solar Cells

Third-generation solar cells are single junction solar cells which are capable of surpassing the Shockley–Queisser limit of 31–41%.

1.3.3.1 Dye-Sensitized Solar Cells (DSSC)

It is a photoelectrochemical cell encompassing a photosensitizer dye molecule and a/ an redox mediator/electrolyte placed between a semiconductor electrode and a counter electrode. The working principle of DSSCs is like mimicking nature's photosynthesis. The DSSCs are made up of 5 components, a transparent conducting film, a semiconductor layer, a sensitizer, an electrolyte, counter electrode. The sensitizer layer absorbs sunlight and produces electron–hole pairs, it is then injected into the semiconductor layer. Dye molecules do the job of absorbing solar energy. Leaves, seeds, fruits, and flowers which naturally contain pigments like chlorophyll, anthocyanin, flavonoid, and carotenoid act as sensitizers. Synthetic sensitizers can be also used, which may increase the stability and efficiency of DSSC but make the device expensive and toxic. Another way of increasing its efficiency is by using Co-sensitization technique, in this different sensitizers having different absorption regions are mixed so that a wide range of solar spectrum can be absorbed. Ruthenium (Ru), Iridium (Ir) and Osmium (Os) are some of the metal complexes that are commonly used in inorganic dyes [4].

It can work in both indoor and outdoor light conditions. Therefore, it can be used to transform both artificial and natural light into electricity. The advantages of DSSCs include their cost-effectiveness and compatibility with regular roll-printing techniques for fabrication. Additionally, their semi-flexible and semi-transparent properties make them ideal for applications where glass-based systems are unsuitable. But, some of the raw

1.3 Classification of Solar Cells

materials like ruthenium and platinum are expensive. Also, using liquid electrolytes is not suitable for all weather conditions. And mostly, DSSC's power conversion efficiency is typically lower than the top-performing thin film-based solar cells [5].

1.3.3.2 Quantum Dot Solar Cells (QDSC)

In quantum dot solar cells, quantum dots (QDs), a semiconductor particle in nanometer scale is used as the sensitizer to absorb the solar energy from the sunlight. Being on a nanometer scale, its optical and electrical properties vary drastically from the bulk materials. By changing the shape, size, and material of QDs, frequency tuning of the light emitted by the QDs is possible. QDSCs are derived from DSSCs where the QDs replaces the photosensitizer dye molecules. Quantum dots also known as artificial atoms, exhibit certain characteristics similar to those of natural atoms, such as singularity, discrete energy levels, and a bound structure. QDSCs can be synthesised in various methods like Microwave irradiation, laser ablation, chemical ablation, hydrothermal/solvothermal, and electrochemical carbonization. CdS, PbS, and CdSe are some of the QDs used. The advantages of these solar cells can be listed as (i) tunable bandgap based on the size of QDs, (ii) multiple exciton generation (MEG) on the absorption of single-photon, (iii) a high value of extinction coefficient, and, (iv) stability towards oxygen and water [4]. Because of tunable bandgap and higher theoretical efficiency of 44%, large photon absorption is possible for QDSCs [4].

1.3.3.3 Organic Solar Cells

Organic solar cells use organic semiconductors or conductive organic polymers as photosensitisers. Bilayer architecture and bulk heterojunction architecture are the two possible organic solar cell structures. The diffusion length of these organic molecules is so small that bulk heterojunction architecture is preferred. The advantages of these solar cells include high optical absorption coefficient, low fabrication cost, the possibility of combining the qualities of plastic with the properties of semiconductors, and tunability of properties with flexible methods of synthesis [4]. The organic solar cell has an added advantage that it can be prepared by less expensive methods like inkjet printing as well as thermal evaporation. In the case of OSC, it exhibits an efficiency of 11.1% but it is much lesser in the module level. In OSC, unlike other PV technologies, a heterojunction is formed between the donor and acceptor materials to enable efficient charge separation. The working of OSC includes three important processes. (i) Photon absorption, (ii) Charge separation, (iii) Exciton diffusion. When a photon is absorbed by organic materials it will create excitons, then these excitons move towards the donor-accepter interface. However, the organic layer should be thin due to its lower diffusion length. After the excitons reach the interface, they will dissociate to free holes and electrons and then be transferred to the corresponding electrodes. Apart from these advantages, like in the case of DSSCs and PSCs, the stability of the device is relatively low [4].

1.3.3.4 Perovskite Solar Cells

Perovskite is actually the name assigned to the mineral $CaTiO_3$, after L.A. Perovski, a Russian mineralogist [10]. It is then and now used for any compound having a crystalline ABX_3 structure. Here A and B represent cations and X as anion. The most common perovskite solar cell type is the hybrid organic lead perovskite solar cell, where A is a monovalent organic cation, B is Pb(II) and X is an anion of halogen (Cl^-, Br^-, I^-). This perovskite material acts as a sensitizer. The most used perovskite photosensitizers are methylammonium lead trihalide ($CH_3NH_3PbX_3$, where X is an ion of halogen such as Cl^-, Br^-, I^-). Formamidinum lead trihalide ($HC(NH_2)_2PbX_3$) is also a promising perovskite material, with the added advantage of a lower bandgap that is closer to the ideal bandgap for efficient solar cell operation. The fabricating process of perovskite solar cells is also simple. One disadvantage is the use of lead, but an alternative solution is already available, where tin (Sn) replaces lead [5]. In a few years, the efficiency of PSC has improved significantly than other photovoltaic technologies. The drawbacks of PSCs include their instability, with factors such as moisture, oxygen, and UV exposure shortening their lifespan [11]. Various methods, such as encapsulation, are effective in extending their lifespan [4].

1.4 Basic Working of Solar Cells

The solar cell (an electrochemical cell) works on the basis of photovoltaic effect where light energy from the sun is absorbed by the device and converted directly to electrical energy [5]. Unlike fossil fuels due to the direct conversion of light to electric energy, it does not produce any toxic substance.

Figure 1.1 represents the basic structure of a perovskite solar cell. A PV cell is the core component of solar energy generation system for turning sunlight into electrical energy [12]. Two basic functions of a solar cell are (1) photocurrent generation and (2) photovoltage generation for the production of electric power [9]. The p-type in the PV cell refers to the holes (positively charged) donated by the acceptor impurity atoms and the n-type refers to the electrons (negatively charged) generated by the donor impurity atoms [12]. The photovoltaic effect is the basis of the functioning of a solar cell. In solar cells the light-to-electrical energy conversion takes place in three steps: (i) absorption of photons and production of electron–hole pairs, (ii) separation of electron–hole pairs, and (iii) Extraction of electron–hole pairs.

When light falls on the semiconductor material, the photons having energy exceeding the bandgap energy (E_g) are being absorbed by the material. When absorbed photon energy ($E = h\nu$) surpasses the threshold bandgap of the material (E_g), an electron is excited from the valence band (E_v) to the conduction band (E_c) of the semiconductor. This excitation leads to the electron–hole pair generation. This transition occurs only when the photon energy crosses the threshold energy or work function ($h\nu_0$) of the semiconductor.

1.4 Basic Working of Solar Cells

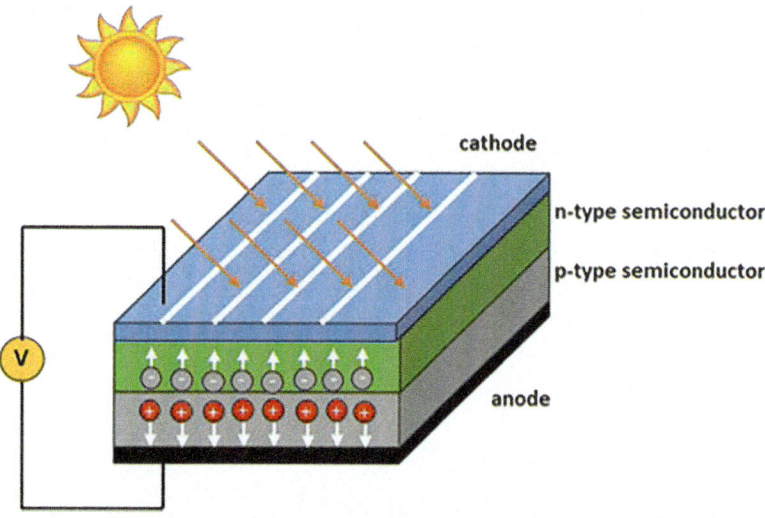

Fig. 1.1 Basic structure of p-n junction solar cells

If the energy of photon exceeds the bandgap, the excessive energy (hv − hv$_0$) is dissipated as heat. (ii) Consequently, the light generated electrons and holes are separated. Electrons and holes move away from the junction through the n region and p region respectively to the metal contact. (iii) The separated electrons and holes are extracted through electrical contacts, generating a transmission of electric current when a load is connected [12].

PV cell exhibits electrical characteristics similar to those of a diode. So under illumination, output current of a solar cell equals the sum of diode current under dark conditions I_d and photocurrent I_{ph} [12].

$$I = I_{ph} - I_d$$

The I-V characteristics of the solar cell under illumination is shown in the Fig. 1.2, which illustrates the relationship between power, voltage, and current. The solar cell must give maximum output power P_{max} at maximum voltage and maximum current I_{max} [12]. The fundamental external parameters of a solar cell are the open-circuit voltage (V_{OC}), the fill factor (FF), the short-circuit current (I_{SC}), and the solar energy conversion efficiency or the power conversion efficiency (η) [9]. where.

I_{SC} is the short circuit current when V = 0 and resistance R = 0
V_{OC} is the open circuit voltage when I = 0 and R = ∞
FF = $I_{max} V_{max} / I_{SC} V_{OC}$ and $P_{max} = I_{max} V_{max}$
The power conversion efficiency PCE of the solar cell is obtained as

$$\eta c = (I_{max} V_{max} / P_{in}) \times 100\%$$

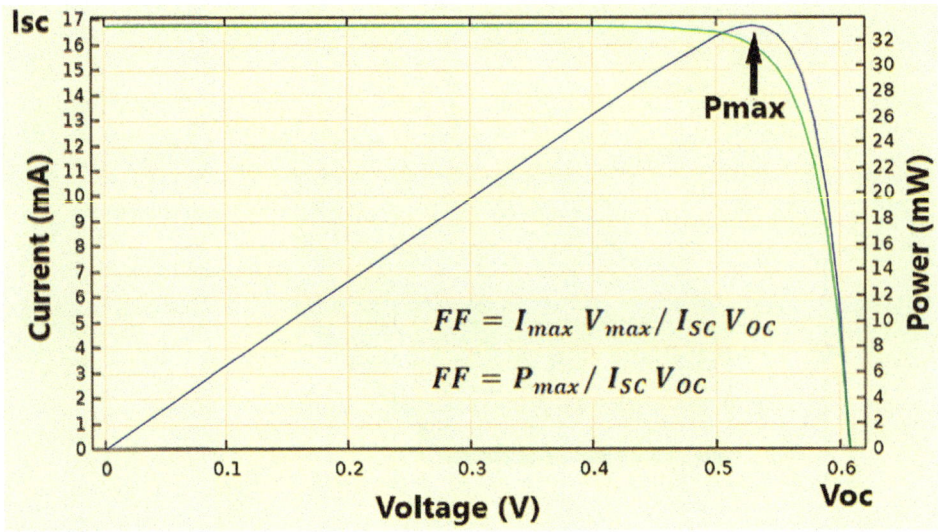

Fig. 1.2 I-V and P-V charecteristics of solar cell (open access)

where P_{in} is the incident power density which is equal to 1 KW/m^2 or 100 mW/cm^2 [12].

In inorganic or organic solar cells, electron and hole separation is a crucial step prior to the recombination. In inorganic solar cells, the electron–hole separation happens due to the intrinsic potential formed at the PN junction. The capability of light generated minority charge carriers to reach the PN junction prior to recombining with the nearby majority carriers is a critical factor in determining the efficacy of inorganic solar cells. In the case of organic solar cells, specific electron and hole transfer mechanisms are followed to transfer the excitons produced to the respective electrodes [4].

1.5 History and Evolution of Perovskite Solar Cells

Third-generation solar cells like organic photovoltaic cells (OPV), DSSCs, and PSCs are advantageous over the first- and second-generation solar technology because they have high-performance efficacy and cheap manufacturing cost. Also, they require only low-temperature processing.

Among these different types, perovskite solar cells have become the focal point in recent years, as they have shown rapid progress in efficiency, increasing from 3.8 to 26.1% in just a decade. Figure 1.3 shows the efficiency improvement of PSCs over the years compared to other solar cell technologies.

1.5 History and Evolution of Perovskite Solar Cells

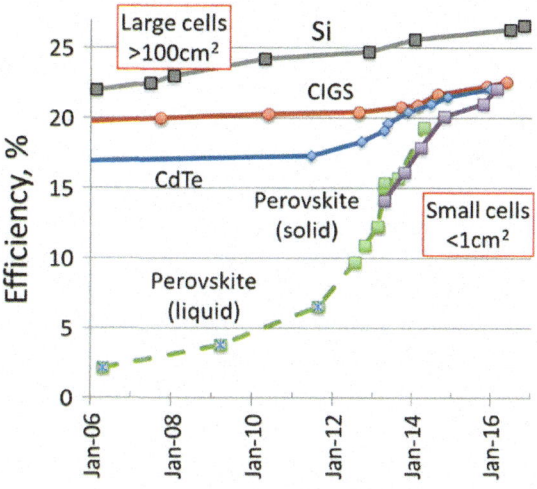

Fig. 1.3 Efficiency versus year graph, growth of perovskite solar cells (Copyright)

It is also particularly attractive for use in tandem solar cells by combining thin films or crystalline silicon solar cells [13, 14]. Organic or inorganic lead halide perovskites have been fabricated and recognised as promising materials for more economic and high-performance photovoltaics. And over the recent years, it has become one of the most exciting developments in photovoltaics. Despite challenges faced regarding the stability of the perovskite and its environmental problems, perovskite dominates in the field of research due to its remarkable potential [15]. The maximum stability achieved is 10,000 h, which is low with respect to the crystalline silicon solar cells. Additionally, many of the high-performance PSCs reported in scientific studies are fabricated on areas as small as 0.01 cm^2; the efficiency tends to decrease as the area increases [13]. As for perovskites solar cell, hybrid organic–inorganic metal halide perovskites like methylammonium lead halides [MAPb(I,Cl,Br)$_3$], and formamidinium lead halides [FAPb(I,Cl,Br)$_3$] are the most widely recognized compounds which exhibits perovskite structure. These materials show photovoltaic, ferroelectric, and pyroelectric properties. They are plentiful, have simple processability, and are well suited for large-scale solution processing techniques, including roll-to-roll printing [16]. Most importantly, they possess a large absorption coefficient, tunable bandgap, and small exciton binding energy [17]. Additionally, their characteristics, such as flexibility, low weight, and semi-transparency, make them even more promising [13, 18].

The notion of perovskite solar cells was arised from DSSCs. Miyasaka and co-workers replaced the liquid sensitizer of the DSSC with solid-state dye. In 2009, an efficiency of 3.8% was gained using methylammonium lead iodide (MAPbI$_3$), an organic–inorganic hybrid perovskite, as the light sensitizer in a photoelectrochemical solar cell [19]. In 2011, Im et al. obtained a more improved efficiency of 6.5% by using quantum dot-sensitized MAPbI$_3$ nanocrystals. In 2012, Kim et al. achieved a ground-breaking advancement in perovskite photovoltaics by incorporating MAPbI$_3$ nanoparticles into solid-state mesoscopic PSCs, resulting in an energy conversion rate of 9.7% and air stability for 500 h with no encapsulation. As for meso-super structured solar cells (MSSC), the efficiency increased to 10.9%. Subsequently, the Snaith group pointed out that a nanostructure is not essential for achieving high PCE, attaining 15% efficiency with a simple planar structure. The efficiency was further enhanced to 19.3% through interface optimization and ETL doping. Using the Intramolecular Exchange Process (IEP), a PCE of over 20% was gained in FAPbI$_3$-based PSCs. Yang et al. further enhanced the efficiency to more than 22% by optimizing the dripping solution to control iodide ion deficiency. A PCE of 23.7% was recently achieved in PSCs using FAPbI$_3$ doped with methylenediammonium dichloride (MDACl$_2$) and a double halide layered architecture (DHA) comprising a superthin wide-bandgap perovskite over a narrow-bandgap layer prior to the hole transporting material (HTM) [13].

When light enters through the transparent cathode, photons with energy exceeding the bandgap of the semiconductor are absorbed. After absorption, electrons from the valance band will get excited to the conduction band and electron–hole pairs will be created based on the binding energy between electrons and holes. The cell's intrinsic potential of the cell separates the electron–hole pairs and will get transported to positive and negative terminals respectively.

1.6 Drawbacks of Perovskite Solar Cells

The major drawbacks of PSCs include.

- Degradation of the perovskite layer
- Lack of perovskite stability
- Toxicity of lead-based PSCs.

The degradation effect of PSC is one of the critical factors which affects the performance of PSC. Perovskite materials made up of metal halides, such as MAPbI$_3$, and FAPbI$_3$, exhibit remarkable efficiency and photovoltaic performance. However, these systems are affected by stability issues. Environmental factors like heat, humidity, and UV exposure will significantly affect the performance of PSC and gradually degrade the perovskite materials. Due to several degradation factors, the stability of PSCs is much lesser, which is a major critical aspect of the commercialization of PSCs. Methods like encapsulation, material engineering, and addictive engineering are some of the ways to increase the stability of PSC. However many years of research are needed to address these issues before this technology can become suitable for commercialization.

Toxicity is another critical issue in the field of photovoltaics. The most efficient PSCs use lead-based perovskite materials as active layers. Even though lead is used on a much smaller scale in PSCs compared to lead-based batteries, it still needs to be addressed. Tin-based perovskites are considered an alternative to lead-based materials, but their efficiencies are still significantly lower than those of lead-based devices. Another way of reducing lead hazards is recycling, life ended solar modules must be disposed of carefully to avoid contamination to the environment. Finding a material that has higher efficiency, stability and less toxicity is the key to commercialising perovskite solar cells [20, 21].

References

1. Y. Li, Design and synthesis of perovskite materials for photocatalytic and photovoltaic applications (n.d.)
2. U. Shahzad, The need for renewable energy sources. ITEE J. (n.d.)
3. Li, X. Development of highly efficient organic and perovskite solar cells (Doctoral dissertation), Nanyang Technological University (2018)
4. M.V. Dambhare, B. Butey, S.V. Moharil, J Phys Conf Ser (2021)
5. A. Mohammad Bagher, Am. J. Opt Photonics **3**, 94 (2015)
6. N. Rathore, N.L. Panwar, F. Yettou, A. Gama, Int. J. Ambient Energy **42**, 1200 (2021)
7. D. Suthar, S. Chuhadiya, R. Sharma, Himanshu, M.S. Dhaka, Mater. Adv. **3**, 8081 (2022)
8. J. Ramanujam, A. Verma, B. González-Díaz, R. Guerrero-Lemus, C. del Cañizo, E. García-Tabarés, I. Rey-Stolle, F. Granek, L. Korte, M. Tucci, J. Rath, U.P. Singh, T. Todorov, O. Gunawan, S. Rubio, J.L. Plaza, E. Diéguez, B. Hoffmann, S. Christiansen, G.E. Cirlin, Prog. Mater. Sci. **82**, 294 (2016)
9. Adeyinka, Adekanmi M., Onyedika V. Mbelu, Yaqub B. Adediji, and Daniel I. Yahya. Int. J. Energy Power Eng 17, 1 (2023)
10. M.V. Dambhare, B. Butey, S.V. Moharil, J. Phys. Conf. Ser. **1913**, 12053 (2021)
11. S.S. Kahandal, R.S. Tupke, D.S. Bobade, H. Kim, G. Piao, B.R. Sankapal, Z. Said, B.P. Pagar, A.C. Pawar, J.M. Kim, R.N. Bulakhe, Prog. Solid State Chem. **74**, 100463 (2024)
12. A.S. Al-Ezzi, M.N.M. Ansari, Appl. Syst. Innov. **5** (2022)
13. S. Khatoon, S. Kumar Yadav, V. Chakravorty, J. Singh, R. Bahadur Singh, M.S. Hasnain, S.M.M. Hasnain, Mater. Sci. Energy Technol. **6**, 437 (2023)
14. F. Cao, L. Bian, L. Li, Energy Mater. Devices **2**, 9370018 (2024)
15. M.A. Green, A. Ho-Baillie, ACS Energy Lett. **2**, 822 (2017)
16. M.L. Petrus, J. Schlipf, C. Li, T.P. Gujar, N. Giesbrecht, P. Müller-Buschbaum, M. Thelakkat, T. Bein, S. Hüttner, P. Docampo, Adv. Energy Mater. **7**, 1700264 (2017)
17. M. Hadadian, J.-H. Smått, J.-P. Correa-Baena, Energy Environ. Sci. **13**, 1377 (2020)
18. N.G. Park, J. Phys. Chem. Lett. **4**, 2423 (2013)
19. C.-G. Wu, C.-H. Chiang, Z.-L. Tseng, Md.K. Nazeeruddin, A. Hagfeldt, M. Grätzel, Energy Environ. Sci. **8**, 2725 (2015)
20. Mohammad, Ashif, and Farhana Mahjabeen. BULLET: Jurnal Multidisiplin Ilmu 2, . 5 (2023)
21. S. Thomas, A. Thankappan, Perovskite photovoltaics (2018)

Perovskite Materials

2.1 Structure of Perovskites Material

Gustav Rose, a Russian scientist, made the initial recognition of perovskite in 1839. The name comes from Lev Perovski's additional research on this mineral, which is composed of calcium titanium oxide ($CaTiO_3$). Researchers show a huge interest in perovskite materials because of their availability in nature. Apart from solar cells, it is also used in superconductivity, ionic conductivity, magnetoresistance, and other telecommunication and microelectronic fields. Perovskite materials are capable of generating more electron–hole pairs from the same quantity of light as compared to other photovoltaic materials. The perovskite structure is portrayed in the figure. The perovskite material's arrangement can be represented as ABX_3, where X is a halide anion, B is a divalent cation, and A is a monovalent cation.

The perovskite structure lies in the form of ABX_3, in the case of PSCs A site represents organic cations like MA (Methylammonium) and FA (Formamidinium). The B is a divalent cations like Cu^{2+}, Mn^{2+}, Ge^{2+}, and Pb^{2+} while X represents negatively charged halogen like F^-, Br^-, and Cl^-. In general, for the ABX_3 framework generally A site is larger than that of B site. In the case of PSCs, $MAPbI_3$, $FAPbI_3$, and $MAPbBr_3$ are the most commonly used active layers that have perovskite structure. These materials also exhibit photovoltaic, ferroelectric, and pyroelectric properties. $MAPbI_3$ exists in different phases at different temperatures. It exists in orthorhombic structure at low temperatures of − 113 °C, tetragonal structure at − 113 to 57 °C and pseudocubic above 57 °C. The material $MAPbI_3$ have a bandgap value of 1.53 eV while $FAPbI_3$ exhibits a lower bandgap of 1.45 to 1.51 eV [1]. The material's thermal stability gets better when the MA cations are swapped out for much larger FA cations. However, due to the larger radius of FA cations, the energy barrier for the intercalation between FA and PbI_6 octahedral network also increases [2]. Thus high annealing temperature of 150 °C is necessary for the perovskite formation of FAPbI$_3$, while MAPbI$_3$ only needs 90–110 °C. The key drawback of FAPbI$_3$ is that, at room temperature, its cubic alpha phase is unsteady and transforms gradually into a photoinactive delta phase with a large bandgap value [2]. Apart from that lead halide materials like MAPbI$_3$ and FAPbI$_3$ exhibit higher absorption coefficients, large diffusion length, and direct bandgap which is suitable for solar cell applications [1, 3].

2.2 Major Perovskite Material

Perovskite material serving as an active layer in PSCs has an ABX$_3$ structure composed of different organic or inorganic cations. Figure 2.1 represents the structure of perovskite material. The A site is made up of monovalent cations like $CH_3NH_3^+$ methylammonium (MA), $CH(NH_2)^{2+}$, and formamidinium (FA); Rb^+; and Cs^+ [4]. Divalent cations such as Sn^{2+}; Pb^{2+} make up the B site. Halide anions comprising Br^-; I^- Cl^- and so forth are composed of the X site. Figure 2.2 indicates the schematic of the perovskite structure ABX$_3$, illustrating the positions of the A, B, and X lattice sites [5]. MAPbI$_3$ is one of the materials which are used intensively for photovoltaic activities [6, 7]. These types of materials which have single ions in these sites are called simple perovskites. For MAPbI$_3$ the Goldschmidt's tolerance factor, 't' is nearly 0.91 and an ionic radius of 0.132 nm for Pb^{2+}, 0.206 nm for I^- and 0.18 nm for MA^+ suggesting a tetragonal phase [8, 9]. According to earlier research, the MA^+ cations occupy the octahedral interstices, and the PbI$_6$ octahedra in MAPbI$_3$ were corner-connected [10].

Although these MA^+ cations are not directly involved in the formation of the band structure, they increase structural stability by compensating for charges within the PbI$_6$ octahedra, which is mostly accomplished through electrostatic interactions. For MAPbI$_3$

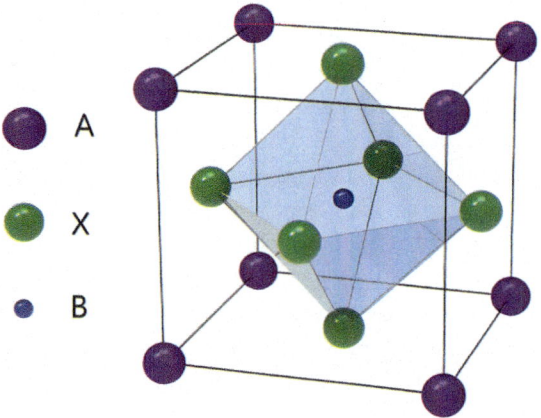

Fig. 2.1 Perovskite structure (open access)

2.2 Major Perovskite Material

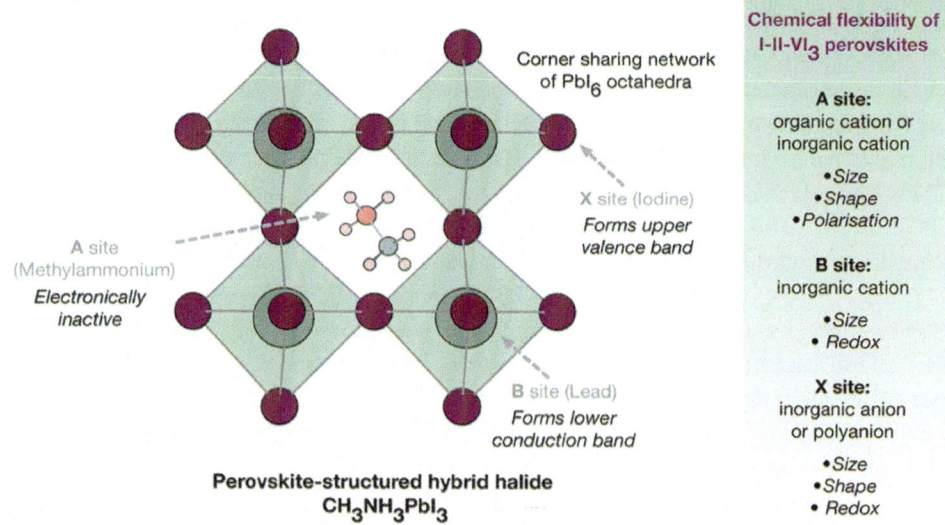

Fig. 2.2 Perovskite structured hybrid halide (copyright)

a wide bandgap of 1.5–1.61 eV is reported, but based on the Shockley-Queisser limit, the optimal bandgap of an ideal single junction solar cell is in the range of 1.1–1.4 eV. Thus several researches are going on to decrease the bandgap value to near ideal bandgap value. Another problem that is facing the researchers is the MAPbI$_3$ shows a phase change from tetragonal to cubic at 54 °C, which is within the solar cell operation temperature range [11]. This is one of the reasons which stops its commercialization. FAPbI$_3$ is a material that exhibits no phase transition in the range of 25–150 °C [12]. Investigations were carried out to further strengthen the optoelectric behaviour of MAPbI$_3$ by replacing MA$^+$ with cations like Cs$^+$, Rb$^+$, and FA$^+$ [13, 14].

2.2.1 MA$_x$FA$_{1-x}$PbI$_3$ Based System

Changing the size of a cation will influence the optoelectric properties of the material. The lattice will expand or contract if A cations with larger (FA$^+$ = 0.19 to 0.22 nm) or smaller (Cs$^+$ = 0.167 nm) ionic radii mix together; this will impact the B-X bond length and could affect the material's bandgap [8]. The mixed perovskite of methylammonium and formamidinium were prepared in a two-step deposition method, firstly dipping or spin coating the preprepared PbI$_2$ with MA$_x$FA$_{1-x}$I solution in IPA. For the composition MA$_{0.6}$FA$_{1-x}$PbI$_3$ the acquired PCE is around 13.4%, and shows an energy bandgap value of E$_g$ of 1.53 eV. Various works were reported in MA$_x$FA$_{1-x}$PbI$_3$ inverted perovskite system using different deposition techniques like spin coating, CVD (Chemical vapor deposition), and LP-VASP (low-pressure vacuum assisted process) having high film quality. By adjusting the system's composition to $MA_{0.6}FA_{0.4}PbI_3$, the PCE rises further to 18.3%. When Cu was used as the top electrode in this system, the performance of the device further increased to a PCE of 8% for 1 cm^2 and 16.48% for 0.09 cm^2 active area.

For a standard architecture of $MAPbI_3$ system, a PCE of 20% is reported while for an inverted structure it is reduced to 18.1%. However, the $FAPbI_3$ prepared by advanced techniques like solvent engineering, organic-cation displacement and $HPbI_3$ precursor shows a PCE of 13.5–18%, which is lower than $MAPbI_3$. Since the ionic radius of FA^+ cations is larger compared to that of MA^+, the bandgap of $FAPbI_3$ is anticipated to be between 1.47 and 1.55 eV. Compared with $MAPbI_3$ a higher PCE is expected for the material $FAPbI_3$, since its absorption range extends to near-infrared. The $FAPbI_3$ shows more short circuit current (Jsc) compared to $MAPbI_3$. However, the reason for less PCE is its poor fill factor (FF) [15]. Another limitation of the $FAPbI_3$ system is its less stability at ambient temperature. It exists in two polymorphs, (i) δ-$FAPbI_3$ of hexagonal phase (ii) α-$FAPbI_3$ of trigonal phase. The material slowly decomposes to photoinactive δ-$FAPbI_3$ at room temperature, which will reduce the stability of the device [16]. Studies found that incorporating MA^+ ($MA_xFA_{1-x}PbI_3$) into the perovskite structure of $FAPbI_3$ will enhance its stability. Due to the less ionic radius of MA^+ as compared to FA^+ cations, it exhibits stronger interaction with PbI_6 which leads to 10 times more dipole moment. Thus incorporating MA^+ cations into the perovskite structure will increase the stability of the $FAPbI_3$ material [13].

2.2.2 $Cs_xFA_{1-x}PbI_3$ System

For this type of system, the Cs quantities from 0.1 to 0.2 show an increase in PCE value compared to the pure $FAPbI_3$ system. With the ordinary PSC structure (FTO/TiO$_2$/Perovskite/spiro-MeOTAD/Au), it portrays PCE ranging from 5 to 15% [17]. It is also reported that this system shows higher stability compared to pure $FAPbI_3$. The encapsulated system is stable up to 250 h with continuous illumination while unencapsulated will last to 350 h in less than 15% humidity [18, 19]. The partial substitution of FA^+ with Cs^+ will increase the interaction with FA^+ and Iodine due to the contraction of cubooctahedral volume. The lattice parameters of $Cs_xMA_{1-x}PbI_3$ were also calculated and obtained a value of 6.363 Å and 6.310 Å for x = 0 and x = 0.25 respectively.

Similar to $FAPbI_3$, $CsPbI_3$ also shows polymorphism from the alpha to delta phase, but the phase transition occurs at a higher temperature of greater than 300°C. They exhibit an energy bandgap value of 1.67–1.73 and show a low PCE of less than 2.9% [16]. However, mixing small quantities with $FAPbI_3$ will enhance the stability of the material. It has another advantage that since the δ to α transition occurs at 125–165 °C, mixing Cs in the ratio of 45% will decrease the transition temperature to room temperature [13, 19].

2.2.3 $Rb_xFA_{1-x}PbI_3$ System

Recent studies show that Rb mixed PSC shows more stability and efficiency than Cs-based PSC. The Rb^+ has an ionic radius of about 0.152 nm, but this material is rarely investigated since it mostly exists as δ-$RbPbI_3$. It exists only in δ phase and the bandgap of the material is between 2.7 and 3.1 eV [20]. The $RbPbI_3$ doesn't show a black α phase in its temperature cycle, it shows yellow colour at 28 °C and it even exists in yellow at 380 °C while in the case of Cs based system, it will change to a black α phase. If we heat it further both the $RbPbI_3$ and $CsPbI_3$ will melt and stop showing black α phase [20]. Studies showed that only a minimal amount of Rb of X < 0.05 can be added to $Rb_xFA_{1-x}PbI_3$ or it will form phase segregation. DSC studies show that there is a temperature difference of 10 °C observed for δ to α phase compared to pure $FAPbI_3$. The transition from δ to α phase is faster for $Rb_{0.05}FA_{0.95}PbI_3$ (~ 60 s) compared to $FAPbI_3$ (~ 4 min). More importantly, the efficiency of the device raised from 14.9 to 16.2%, and its stability against humidity also increased. In pure $FAPbI_3$ the films start showing degradation to δ phase within 6 h when it is exposed to 85% of relative humidity at 25 °C. Cs-based $FAPbI_3$ shows slightly higher stability while Rb-based devices doesn't show any absorption range even if it is exposed to 25 h. Both the Rb^+ and Cs^+ have similar properties electronically and chemically, the only difference is the slight variation in their ionic radius, the Cs^+ ions have an ionic radius of 0.167 nm and Rb^+ have 0.152 nm [12]. Despite the unnoticeable disparity in ionic radius, $Rb_{0.05}FA_{0.95}PbI_3$ reveals greater stability than $Cs_{0.05}FA_{0.95}PbI_3$. Rb-based solar cells exhibited an efficiency of > 97% compared to its initial efficiency, for a long-term period of 1 month [9, 20].

2.2.4 $MAPbI_{3-x}Br_x$ System

Substituting Iodine with Br in $MAPbI_3$ can tune the bandgap of the system effectively. It was reported that by the solvent engineering method, researchers reported an efficiency of 12.3% for x = 0.10 and 16.2% for x = 0.15. Later different preparation methods were considered such as inkjet printing, and LP VASP (low pressure vapor assisted solution). Due to a smaller bandgap of $MAPbI_3$, a Br content of less than 10% in the system gives the highest initial efficiency while a higher Br content of > 20% results in better stability against humidity (Relative humidity of 55%). This was associated with a change in structure from tetragonal to pseudocubic at x = 0.13, which results from a higher tolerance factor (t) brought on by Br's smaller ionic radius [9]. Vegard's law employs to mixed perovskites, which are made up of two distinct perovskites that have identical lattice constants. The formula suggests the linear relationship of the lattice parameter with composition when there exist no significant electronic effects [9]. A minor variation from

Vegard's law was noted in the cubic regime of x > 0.57, which designates more interactions in the mixed halogens [9]. Moreover, they observed a blue shift of the absorption edge, which is a quadratic dependence on the Br content and indicates an increase in the bandgap created by Br [9].

2.3 Comparison with Traditional Solar Cell

PSCs are among the more recent contenders in the arena of photovoltaics. PSC belongs to the group of third-generation solar cells that are considered more economical, of higher efficiency and more environmentally friendly. Up to 25% efficiency has been shown by the conventional single junction silicon solar cell. The multijunction GaAs solar cell possesses 40% efficiency, whereas single crystal GaAs solar cell showed an efficiency of 29%. Quantum dots and DSSCs are other kinds of solar cells that are renowned for their easy design and affordable manufacturing. After years of research, it gained an efficiency of 13%, and it is not yet commercialized due to its lower efficiency and poor long-term performance.

The importance of PSCs is its fast growth in efficiency in a short period. It was first introduced in 2009, which exhibits an efficiency of 3.8%. After a few years of research, the efficiency is increased to 20% [13]. It is much higher than quantum dots solar cells. Currently, the efficiency of PSC has reached up to 26%.

Presently, three wafer-based solar cell exists

(i) Crystalline Si solar cell
(ii) Gallium Arsenide (GaAs) solar cell
(iii) Multijunction solar cells (III–V).

2.3.1 Crystalline Silicon Solar Cell (c-Si)

Crystalline silicon solar cells are the most popular and commercial-level solar cells available today. Its higher efficiency, stable performance and non-toxic behaviour make it more popular. It also possesses certain disadvantages, it shows an indirect bandgap of 1.1 eV, which makes it a poor light absorber. This drawback was overcomed by using thick and brittle wafers to absorb more incident light, making it difficult to fabricate flexible solar cells. But for Si solar cells a high level of material purification is needed, which increases the overall cost. Apart from these disadvantages more than 90% of module production globally is based on silicon solar cells. This type of solar cell requires a high level of material purification, which is very expensive. Current research uses thin membranes as starting materials instead of wafers. By this method, silicon solar cells can be constructed using low-cost materials making the entire solar cell more economical. Even though the efficiency is slightly lesser this method is still considerable due to its cost of production. The scalability of large modules is still unknown and unproven [13].

2.3.2 Gallium Arsenide (GaAs)

On a global scale among the 40 GW of total solar energy produced most of them are from Si solar cells. Even though the production cost of Si solar cells is reduced to some extent, the increment in efficiency is still poor [21]. This makes the researchers to find out a new material that is more efficient for solar cells. GaAs is such a material that shows higher efficiency. This type of solar cell is mostly used for space purposes.

Apart from the increased cost of production, it has various technological advantages compared to silicon solar cells, which includes its higher electron transfer in crystalline structure. It is a III–V single junction semiconductor that has higher absorption coefficients, shows direct bandgap, low non-radiative energy loss etc. All these make it a perfect solar photovoltaic device. Due to the direct bandgap of 1.42 eV, it exhibits a PCE of 28.8%. Making it more suitable for multi-junction solar cells.

Compared to most commonly used Si solar cells, it has many advantages:

- It has wide coverage of the solar spectrum
- Direct bandgap
- High photoconversion efficiency
- High radiation resistance.

The cost of production of GaAs cells is more than 10 times that of Si solar cells. Lowering the production cost is a difficult task since the overall efficiency is greatly dependent on the imperfections in the crystal and impurities. This limits its operation for space communications or other higher-efficiency applications [13, 22].

2.3.3 Multi-junction Solar Cells (III–V)

The GaAs single junction can convert light to energy, even above the bandgap value. However, the photons absorbed above the bandgap of the materials cannot produce electron–hole pairs since the energy formed is lost in the form of heat in the lattice itself, which limits its overall efficiency. In Multi-junction solar cells, p–n junction semiconductors of different absorption regions are connected in series. Theoretical research shows that the PCE can be further improved by using multijunction solar cells. It is theoretically reported that the efficiency can be increased up to 72% by making a multijunction solar cell with 32 bandgaps. These solar cell is made by stacking different single-junction solar cells having suitable bandgaps; this method can convert light into energy more efficiently while minimizing heat loss. The efficiency is increased to 30% when researchers use gallium arsenide with gallium antimonide in tandem solar cells; gallium arsenide has a bandgap value of 1.42 eV, while gallium antimonide has 0.72 eV.

The production of multi-junction solar cells is also very limited due to its higher material costs and complex manufacturing process. This is mostly used for space communication purposes due to its higher efficiency and low-temperature sensitivity [13, 23].

2.3.4 Perovskite Solar Cell

A solar cell is a device which directly transforms light to electrical energy. The development of solar cell can be categorised into three Organic solar cell, Si solar cell and thin film solar cells. Even though as already established silicon solar cell has so many advantages, its cost of production limits its wide usablility. The third generation solar cell such as perovskites solar cell have the potential to replace conventional solar cell [24]. It has higher efficiency and lower cost compared to Si solar cell technology [25]. Also, its other characteristics like flexibility, low weight and its semi-transparency make it more promisable. It can be considered an advanced version of DSSC. Since efficiency is the most important factor in solar cells, PSCs exhibit high efficiency due to their larger absorption region and lower bandgap [26]. PSC is widely accepted as a future photovoltaic technology that is economical and more environmentally friendly. However, in some PSCs, lead is used to make the active layer, which is still an issue for large-scale production of PSCs. The PSC technologies can be used with first and second-generation solar cells [27]. Compared to other solar cell technologies, the efficiency of PSC is closer to that of silicon solar cells (~ > 25%) [28]. PSC achieved this efficiency in a short period of time while silicon solar cells took several decades to achieve this efficiency. Material abundance is another advantage of PSC. Even though silicon is abundant in the earth, its purification increases the cost of production. Also, it can be prepared even at low temperatures, making them more suitable for flexible solar cells. Even though it has various advantages, PSCs face several stability issues like sensitivity to UV light, temperature, and moisture content in the atmosphere. Since it is an emerging technology, if these issues were solved in the coming decades, PSC will be a strong opponent to other solar cells which rule the market today [5, 13].

But to commercialise this technology certain challenges have to be overcome. It has a huge disadvantage which is its stability. PSCs are very much prone to degradation due to moisture, UV lights and temperature [29]. Thus, without a proper solution, this can't be commercialised. The lifetime of PSC must be at least half of a silicon solar cell. Several researches have been carried out to increase its stability and to find an alternative material to replace Pb.

PSC can be fabricated through physical and chemical methods which include sputtering, dip coating, spin coating, and chemical vapour deposition [30, 31]. To get maximum efficiency at the device level the perovskite layer must be uniform in terms of thickness, quality and roughness. The photon absorption of the perovskite layer directly depends on the thickness of the layer but the increased thickness affects its charge carrier extraction and results in low efficiency [32].

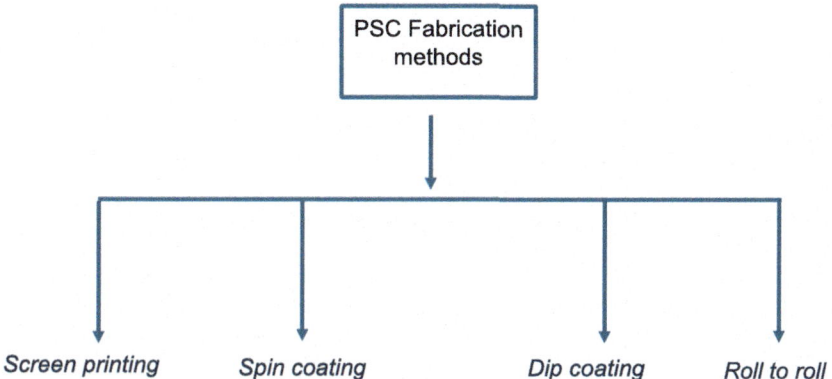

Unlike silicon solar cell which requires an expensive multistep fabrication process that requires higher temperature, high vacuum, and cleanroom for fabrication, the PSC can be prepared at a lower cost without much sophisticated instruments.

References

1. P. Arjun Suresh, G.S. John, A.M. Johnson, U.S. Sajeev, K.V. Arun Kumar, J. Mater. Sci.: Mater. Electron. **35** (2024)
2. P. Wang, Y. Wu, B. Cai, Q. Ma, X. Zheng, W.-H. Zhang, Adv. Funct. Mater. **29**, 1807661 (2019)
3. S. Thomas, A. Thankappan, *Perovskite Photovoltaics* (2018)
4. R. F. Berger, Chem.—Eur. J. **24**, 8708 (2018)
5. N. Ali, N. Shehzad, S. Uddin, R. Ahmed, M. Jabeen, A. Kalam, A.G. Al-Sehemi, H. Alrobei, M.B. Kanoun, A. Khesro, S. Goumri-Said, Int. J. Energy Res. **45**, 19729 (2021)
6. H.J. Snaith, J. Phys. Chem. Lett. **4**, 3623 (2013)
7. N.G. Park, J. Phys. Chem. Lett. **4**, 2423 (2013)
8. J.P. Correa-Baena, A. Abate, M. Saliba, W. Tress, T. Jesper Jacobsson, M. Grätzel, A. Hagfeldt, Energy Environ. Sci. **10**, 710 (2017)
9. L.K. Ono, E.J. Juarez-Perez, Y. Qi, ACS Appl. Mater. Interfaces **9**, 30197 (2017)
10. X. Liu, L. Cao, Z. Guo, Y. Li, W. Gao, L. Zhou, Materials **12** (2019)
11. L.K. Ono, Y. Qi, J. Phys. Chem. Lett. **7**, 4764 (2016)
12. J.W. Lee, D.J. Seol, A.N. Cho, N.G. Park, Adv. Mater. **26**, 4991 (2014)
13. T. Ibn-Mohammed, S.C.L. Koh, I.M. Reaney, A. Acquaye, G. Schileo, K.B. Mustapha, R. Greenough, Renew. Sustain. Energy Rev. **80**, 1321 (2017)
14. W.J. Yin, J.H. Yang, J. Kang, Y. Yan, S.H. Wei, J. Mater. Chem. A **3**, 8926 (2015)

15. Y. Zhou, M. Yang, S. Pang, K. Zhu, N.P. Padture, J. Am. Chem. Soc. **138**, 5535 (2016)
16. C.C. Stoumpos, C.D. Malliakas, M.G. Kanatzidis, Inorg. Chem. **52**, 9019 (2013)
17. Y. Yu, C. Wang, C.R. Grice, N. Shrestha, J. Chen, D. Zhao, W. Liao, A.J. Cimaroli, P.J. Roland, R.J. Ellingson, Y. Yan, Chemsuschem **9**, 3288 (2016)
18. J.W. Lee, D.H. Kim, H.S. Kim, S.W. Seo, S.M. Cho, N.G. Park, Adv. Energy Mater. **5** (2015)
19. Z. Li, M. Yang, J.S. Park, S.H. Wei, J.J. Berry, K. Zhu, Chem. Mater. **28**, 284 (2016)
20. Y.H. Park, I. Jeong, S. Bae, H.J. Son, P. Lee, J. Lee, C.H. Lee, M.J. Ko, Adv. Funct. Mater. **27** (2017)
21. M.T. Hoang, N.D. Pham, J.H. Han, J.M. Gardner, I. Oh, ACS Appl. Mater. Interfaces **8**, 11904 (2016)
22. Adeyinka, Adekanmi M., Onyedika V. Mbelu, Yaqub B. Adediji, and Daniel I. Yahya. Int. J. Energy Power Eng 17, 1 (2023).
23. A. Baiju, M. Yarema, Front. Energy Res. **10** (2022)
24. B. Parida, S. Iniyan, R. Goic, Renew. Sustain. Energy Rev. **15**, 1625 (2011)
25. J. Ciro, R. Betancur, S. Mesa, F. Jaramillo, Sol. Energy Mater. Sol. Cells **163**, 38 (2017)
26. Y. Zhang, Z. Fei, P. Gao, Y. Lee, F.F. Tirani, R. Scopelliti, Y. Feng, P.J. Dyson, M.K. Nazeeruddin, Adv. Mater. **29**, 1702157 (2017)
27. I. Ali, M.R. Islam, J. Yin, S.J. Eichhorn, J. Chen, N. Karim, S. Afroj, ACS Nano **18**, 3871 (2024)
28. J. Huang, S. Tan, P.D. Lund, H. Zhou, Energy Environ. Sci. **10**, 2284 (2017)
29. K.T. Alao, S.I. ul Haq Gilani, T.O. Alao, A.U. Adebanjo, O.R. Alara, Next Energy **7**, 100215 (2025)
30. C.H. Chiang, Z.L. Tseng, C.G. Wu, J. Mater. Chem. A **2**, 15897 (2014)
31. M. Liu, M.B. Johnston, H.J. Snaith, Nature **501**, 395 (2013)
32. D. Liu, M.K. Gangishetty, T.L. Kelly, J. Mater. Chem. A **2**, 19873 (2014)

Fabrication of Perovskite Solar Cell

3.1 Device Structure

Perovskite solar cells mainly consist of 5 layers as shown in Fig. 3.1. The bottom layer of PSC is made of transparent electrodes like ITO (Indium tin oxide) or FTO (Fluorine-doped tin oxide). This layer makes a transparent and conductive layer on the glass plates, which will pass light through it. On top of the transparent layer, there is an electron transport layer (ETL). This layer conducts electrons and blocks holes from the active layer.

The perovskite layer is developed above the ETL, which acts as an active layer in PSCs. This layer produces electron–hole pairs while absorbing light. Above the active layer, hole transport layer (HTL) and finally a metallic electrode is pasted. When illuminated, electrons get excited from the valence band of the perovskite to the conduction band, leaving holes in the valence band [1]. The variation in work function between the ETL and HTL makes the generated electrons and holes to move in opposite directions towards the contacts. This produces current in an external circuit connected between the contacts [2, 3].

These layers individually have vital impact in the overall efficiency of solar cells. Choosing ETL itself is a challenging task, the solar cell becomes significantly efficient if the ETL can extract more electrons from the active layer [4]. This will prevent the recombination of electron–hole pairs. A most commonly used ETL is TiO_2, through research, it is found that adding a compact layer of TiO_2 followed by a mesoporous layer will increase its ability to block holes. This lowers the series resistance and increases the shunt resistance, thereby reducing the overall resistance of ETL and prevents electron–hole recombination. Spin coating, spray pyrolysis or evaporation are the most common methods for coating ETLs in PSCs. Bandgap is the energy difference between HOMO (Highest Occupied Molecular Orbital) and LUMO (Lowest Unoccupied Molecular Orbital). The perovskite material used as an active layer must have a higher absorption region and

Fig. 3.1 Layers of perovskite solar cells

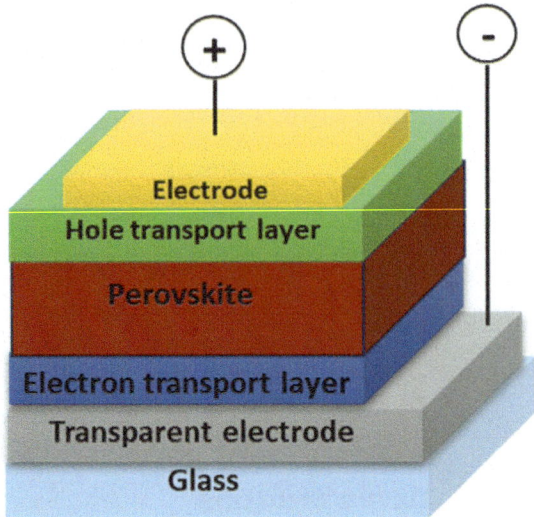

lower bandgap value. A broader absorption region or lower bandgap of perovskite material will increase photon absorption, leading to a higher open-circuit voltage (V_{oc}) [5]. Similar in the case of HTL, a highly conductive HTL shows higher short circuit current density (J_{sc}). Ag and Au are the most widely used conducting electrodes, that can be deposited by vaporising corresponding metals [6].

3.1.1 Electron Transport Layer

In the case of solar cells, the efficiency of the device is greatly dependent on the efficiency of HTL and ETL. Optimising these will increase the open circuit voltage (V_{oc}) and short circuit current (J_{sc}). Materials with a low work function are suitable as the ETL, while materials with a high work function are suitable as the HTL. Compared to the perovskite layer ETL must have the lowest LUMO and highest HOMO level. Since ETL is positioned before the active layer, it must have high transmittance, so that light can reach the perovskite layer. ETL plays a pivotal role in increasing the efficiency of solar cells by extracting and transporting electrons from the active layer and it also acts as a barrier for the conduction of holes. The final performance of solar cells also depends on the electron mobility, trap states, and energy level alignment of PSCs [7].

TiO_2 is the most commonly used ETL in PSCs, but it is observed that the perovskite solar cell that uses TiO_2 as ETL has lower stability compared to TiO_2-free PSC. Using a UV filter blocks UV light, which results in increasing stability over a longer duration. Another drawback is that because of the high annealing temperature of TiO_2 it is not suitable for making flexible solar cells. These drawbacks were recently fixed by incorporating SnO_2 instead of TiO_2. It possesses benefits including low-temperature synthesizing, high electron mobility, deeper connection band, have anti-reflection coating. These properties reduce the charge accumulation on the surface and increase the charge transfer from perovskite to ETL [8].

Rather than metal oxide PSCs organic ETL like PCBM, PEHT, PEHT: PCBM can gain higher efficiency in inverted structures. Many polymers satisfy the condition that the LUMO of the ETL must be lesser compared to perovskite layer. Also to get maximum efficiency the perovskite layer must be of high quality. A smooth, compact perovskite layer having lesser voids depends on the quality of ETL, it will also reduce the unnecessary leakage of current. The substrate's nucleation site decides the crystallization and nucleation behaviour of the perovskite.

TiO_2, an inorganic electron transport material, has a high electron injection rate from the perovskite layer to TiO_2. But as a downside, it also has a higher electron recombination rate. For solving this problem researchers found another material called ZnO. It has a bulk electron mobility of 205–300 cm^2 V^{-1} s^{-1} which is higher than TiO_2. The perovskite solar cell which uses ZnO as ETL shows an efficiency of about 15.7%. It also has another major drawback, which is its lack of stability. Then as an alternate ETL SnO_2 is emerged. It has high transparency, wide bandgap and has bulk electron mobility of 240 cm^2 V^{-1} s^{-1}. Recent studies found that low-temperature processed SnO_2 shows an average efficiency of 13%, while low-temperature sol–gel derived SnO_2 gains a PCE of 16.02% [7].

3.1.2 Hole Transport Layer (HTL)

HTL is vital in extracting the holes and blocking electrons from the active layer of PSC. HTL perform several roles in PSC which are.

- Separation or extraction of the holes produced by the active layer by absorbing light.
- Blocking the electrons' passage to the anode.
- Increasing the stability of the perovskite layer by encapsulating it from the moisture.
- Results in increasing the V_{oc} due to well-matched HOMO energy level.

Spiro-OMeTAD is one of the mostly utilized HTLs, which results in high efficiency at the device level [9]. Even though it has high efficiency, other factors like the high cost of synthesis and lack of stability prevent its commercialization. Perovskite materials like $MAPbI_3$ ($CH_3NH_3PbI_3$) have ambipolar characteristics. However, it is more towards a p-type material than an n-type, this type of cell can be prepared without HTL. Recently an efficiency of 11% is obtained for HTL-free PSCs where the perovskite layer act as HTL as well as active layer. Eliminating the need of separate HTL enables cost effective manufacturing of the device [10]. Besides HTLs like Spiro-OMeTAD, PEDOT:PSS[Poly (3,4-ethylenedioxythiophene): polystyrene sulfonate} and P3HT [poly(3-hexylthiophene-2,5-diyl) are also popular. Also, inorganic HTMs like Cu_2O, CuSCN, CuO, CuI, NiO_x MoS_2 are studied. In general inorganic HTMs exhibit high hole mobility and good stability at lower cost [9, 11].

3.2 Solution Processing Methods

The PSC can be fabricated either by solution based or vapor based deposition technique, as shown in Fig. 3.2. The solution based technique can be further classified into spin coating, roll to roll techniques.

3.2.1 Spin-Coating Method

Spin coating is an efficient and inexpensive method to apply perovskite layers. Figure 3.3 explains the steps involved in the spin-coating technique for fabricating perovskite materials. This method is generally used in the laboratory scale, since the film will be non-uniform in the large area coating. After spin coating, the film is annealed to get good crystalline perovskite structures [12]. By controlling the spin rpm (rotations per minute), acceleration, and spinning time the film thickness can be controlled. This technique can be used for both inverted and non-inverted PSCs [13]. The highest PCE obtained through this method is 22.1%. The difficulty in forming uniform large-scale devices and the issue

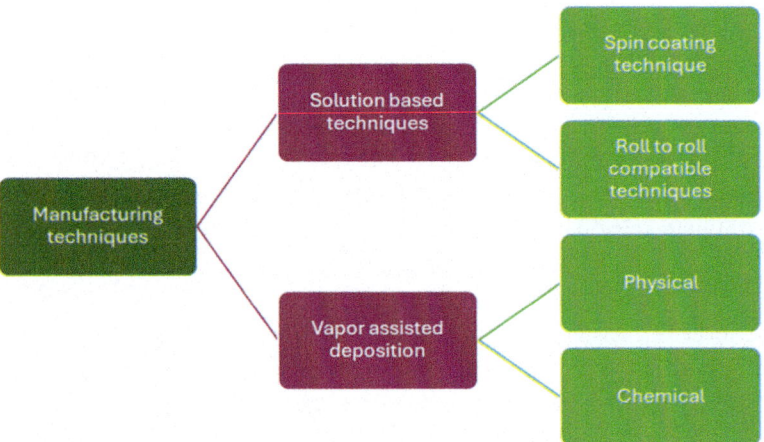

Fig. 3.2 Classification of PSC fabrication technique

3.2 Solution Processing Methods

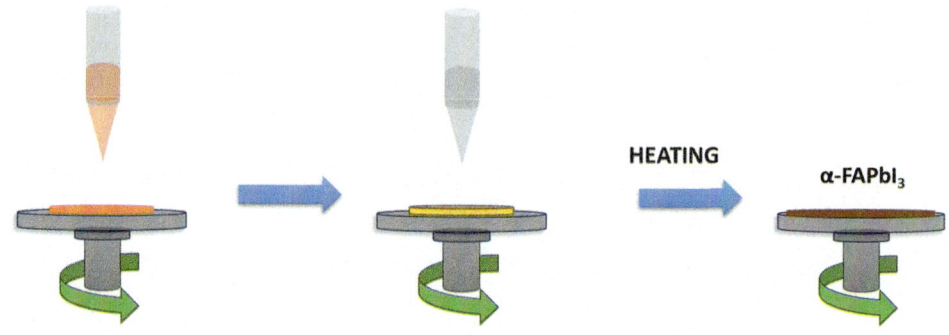

Fig. 3.3 Steps in the spin-coating method

of material wastage limit this method to the laboratory level. The active layer of a PSC can be fabricated either in single step process or two step process. The one-step process involves spin coating the substrate with a precursor solution that contains all the necessary ingredients, such as an organic halide solution like FAI or MAI and a metal halide precursor like PbI_2, and then annealing it to produce the appropriate perovskite material. Primarily, precursor solution is coated onto the substrate, followed by the deposition of organic halides such as MAI or FAI. Finally, the film is annealed to get the desired perovskite material [14].

3.2.2 Drop-Casting Method

Drop casting is an economical technique for synthesising PSCs, it is much identical to the spin-coating method. Figure 3.4 illustrates the schematic diagram of Drop-casting method. The difference of drop casting from spin-coating is the absence of spinning of the substrate. The film thickness depends on the amount of solution dropped into the substrate; other variables include the temperature and annealing time. Volatile solutions are normally preferred for this method, it has an added advantage that there is no wastage of solutions like in spin-coating methods. The difficulty in obtaining uniform films and controlling film thickness are the major drawbacks of this method [14, 15].

3.2.3 Roll to Roll Printing

It is an appropriate technique for producing PSCs. This technique can be used to produce flexible PSCs and can be used to manufacture large-scale devices. Different coating

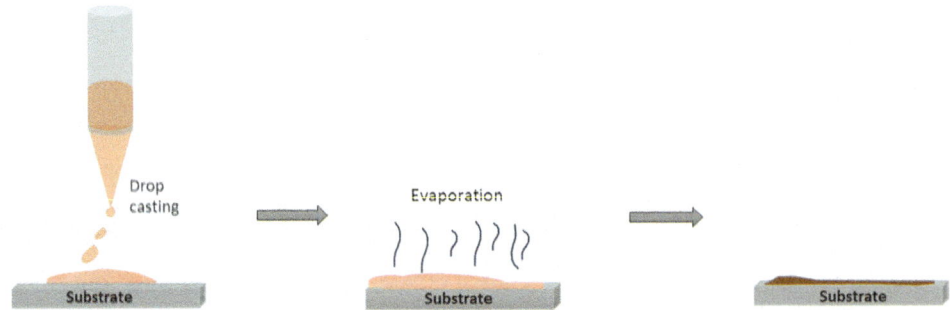

Fig. 3.4 Schematic diagram of drop-casting method

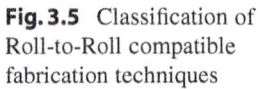

Fig. 3.5 Classification of Roll-to-Roll compatible fabrication techniques

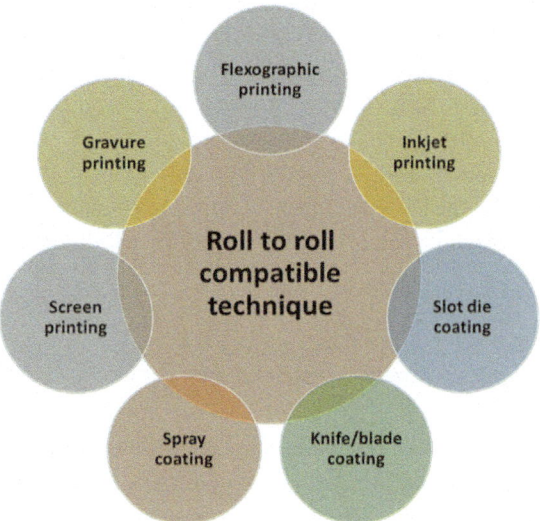

methods can be adopted to produce thin films based on this technique some of them are spray coating, screen printing, blade coating, and inkjet printing. The primary benefit of this process is the ability to create lengthy wafers more quickly and effectively [16] (Fig. 3.5).

3.2.4 Slot Die Coating

This coating technique is generally preferred for fabricating large-area PSCs. This technique can create thin and uniform films directly onto the substrate. Figure 3.6 illustrates slot die coating technique, in this, the liquid material is delivered through the coating head, generally called slot die which is placed above the substrate. The PSCs fabricated

3.2 Solution Processing Methods

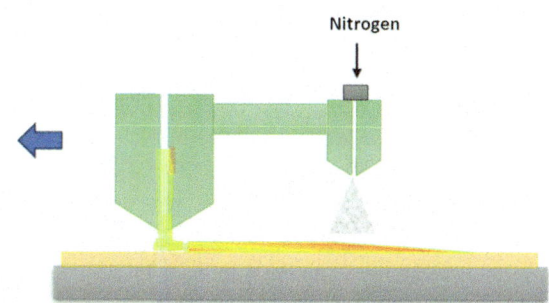

Fig. 3.6 Schematic representation of slot die coating

showed an efficiency of 11% but in the case of large area PSCs of active area, 47.3 cm^2 shows an efficiency of 4.57%. In the case of MAPbI$_3$, for sequential deposition, initially we want to print the PbI$_2$ onto the substrate. This technique can even replace the second step of the sequential deposition technique, which is dip coating or spin coating the substrate using MAI (Methylammonium iodide) solution [16]. It also has the facility of varying the substrate temperature and by this, we can vary the crystallization rate of the perovskite material. The nitrogen air flow provided in this technique helps to form more uniform and pinholes-free films. Like in the earlier discussed blade coating method, regulating the solvent evaporation rate during or after the coating process will enhance the quality of the films [17]. This slot dyeing technique is not only used in PSCs but also in batteries and making flexible solar cells. The material's viscosity and rate of flow will change the quality of the coating process [16].

3.2.5 Spray Coating

This is one of the efficient ways of fabricating efficient flexible solar cells. Compared to the spin-coating technique, which is limited to small-scale devices, this method can be used to produce large flexible solar cells [18]. Inkjet printing, the doctor blade method and the slot die method are other scalable film deposition methods. Among these, the doctor blade method is the cheapest; however, the film deposited by the spray coating method exhibits an efficiency ten times higher than that of the doctor blade method. Figure 3.7 depicts the schematic representation of spray coating method. Compared to spin-coated films the spray-coated films exhibit better optoelectroic properties and have higher thermal stability. This method can be used in both inverted and noninverted PSCs. The highest efficiency obtained through this method is 11%. However, achieving a fully covered and homogeneous film through this method is challenging. This is due to liquid atomization, where the random spraying of liquid droplets onto the film creates patches of varying sizes, increasing series resistance and impacting the overall functionality of the device. To reduce these drawbacks electrostatic spray coating, ultrasonic spray coating, pulsed spray coating, and airbrush pen spray coating are used [14].

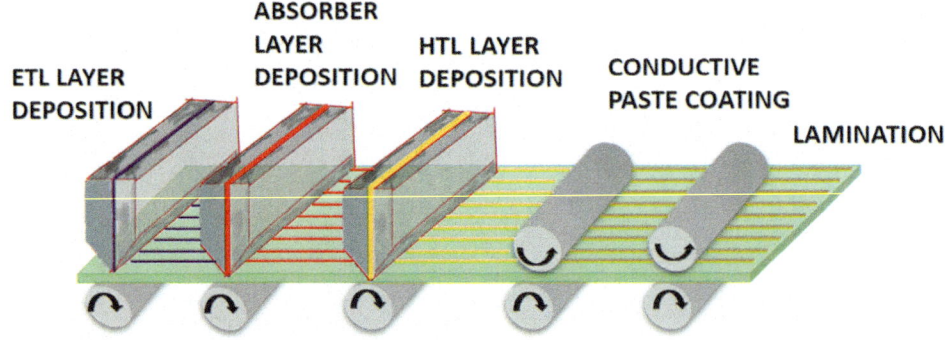

Fig. 3.7 Illustration of spray-coating method

3.2.6 Blade Coating Method

This method can be used to fabricate PSC at the industry level [13]. As shown in Fig. 3.8, it is a simple system that features a screw to adjust the blade height relative to the substrates. Altering the substrate temperature or the airflow over the substrate will change the uniformity of the film. It is found that the blade coating method offers a better film morphology compared to any other techniques [16].

Pinhole formation is an issue that is faced in other techniques, and this will create a non-uniform film deposition. Blade coating is a highly recommended film coating method for developing uniform films. It has been reported that using this method, the perovskite crystallizes at a slower rate, resulting in larger agglomerations, which in turn inhibits

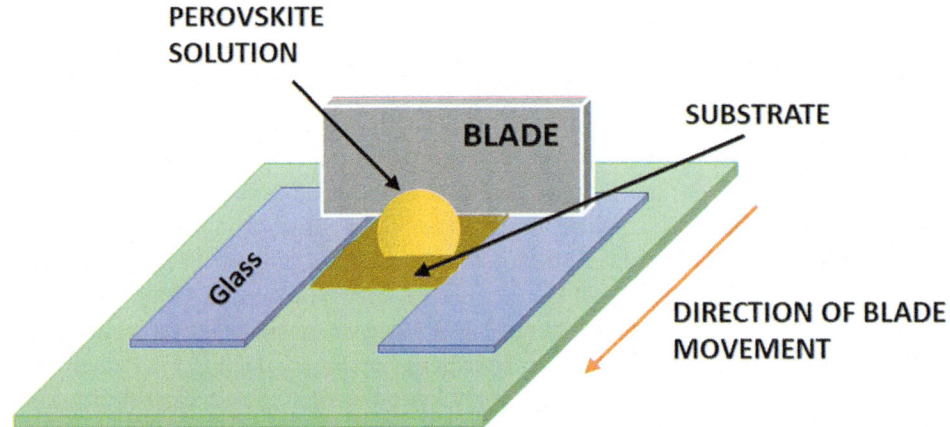

Fig. 3.8 Schematic diagram of blade coating method

moisture permeability to the perovskite layer [19]. Thus this method provides a higher stability and better morphology for the perovskite film [14].

3.3 Vapor-Based Techniques

3.3.1 Vapor-Assisted Deposition

Solution-based deposition and vapor-based deposition are the two techniques used for creating perovskite layers. Vapor-based deposition can be classified into physical and chemical vapor depositions. As shown in Fig. 3.9, precursor coated substrate is exposed to the vapors of secondary reactants. The vapors will react with the precursor material and form the desired phase. For the solution-processed method, the film thickness lies in the range of micrometres, while vapor deposition technique can produce nearly a nanometer range. Additionally, it can produce better uniform and highly crystalline films which results in higher overall efficiency. The highest efficiency obtained by vapor deposition method is 15.4% [18].

In PSCs, the thickness of the film has a considerable impact on its efficiency. If the film thickness is too low, it cannot absorb enough solar radiation from the sun. If it is too thick the time required to reach the electron–hole pairs in electrodes will become

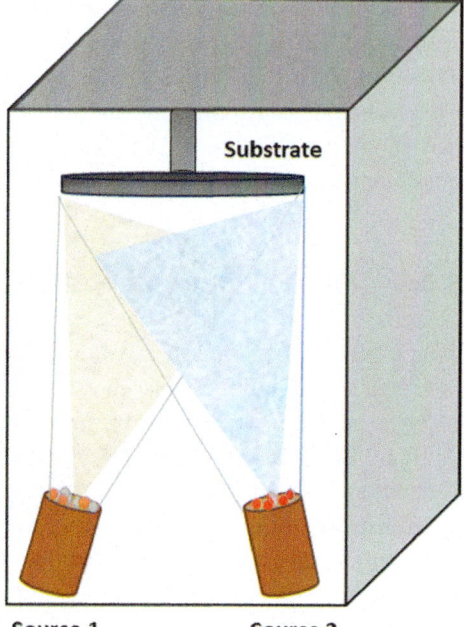

Fig. 3.9 Schematic diagram of vapor-assisted deposition

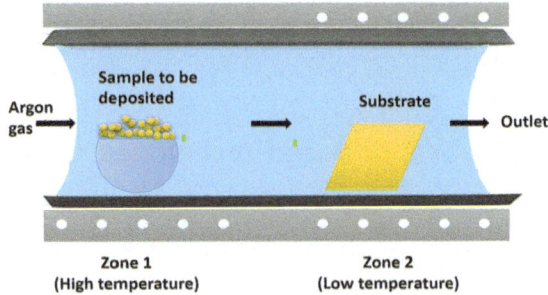

Fig. 3.10 Perovskite film fabricated using chemical vapor deposition

higher, which results in recombination. Additionally, if there are pinholes present in the active layer, they result in direct contact between the HTL and ETL, leading to a lower fill factor (FF) and lower Voc. This type of technique is commonly used for industrial-scale applications, including liquid crystal displays, and solar industry. One of the drawbacks of vapor deposition method is that it requires a vacuum for its process. Applying a vacuum will increase the mean free path of vapors and result in high-quality films [14].

3.3.2 Chemical Vapor Deposition

CVD is one among the reliable vaporization techniques that can be used to prepare large-scale films with fewer pinholes and uniform surfaces with excellent material yield and scalability. The CVD process works by mixing two precursors as shown in Fig. 3.10, which are then transferred along with a carrier gas and deposited onto the substrate. This technique can form uniform, pinhole-free substrates with large grain sizes and long carrier lifetimes. However since vacuum is required for this technique, this technique is not commonly seen in mass production [14].

3.3.3 Physical Vapor Deposition

This is one of the simplest ways of perovskite deposition, which will give good surface coverage and stability against moisture. Figure 3.11 shows the schematic representation of physical vapor deposition. In this method, the material which is in the form of solid or liquid is vaporised into atoms or molecules. These vapors are then passed through a vacuum atmosphere and finally condensed onto the substrate. This method offers better control on the film morphology, thickness and film quality. Unlike other methods like the dual source vapor deposition technique or two-step vapor deposition technique, this will minimise the impurities that are deposited on the substrate and avoids improper heat

3.3 Vapor-Based Techniques

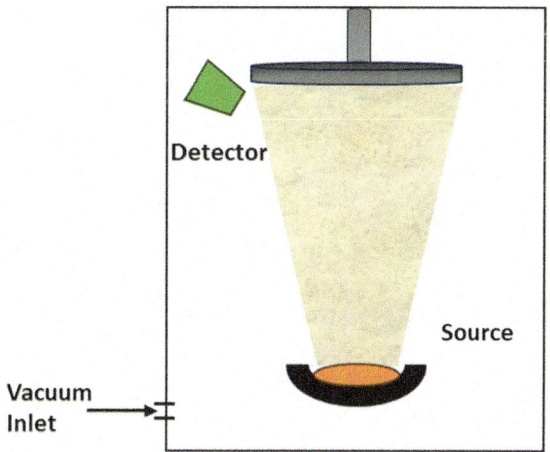

Fig. 3.11 Schematic diagram of physical vapor deposition

treatments. The maximum efficiency attained by this technique in fabricating PSCs is 15.4%. Hence this is an appropriate technique for developing PSCs, this is also used at the industry level as well [14].

3.3.4 Vapor-Assisted Solution Process

Vapor-assisted solution processing can be considered a modified two-step method for fabricating the perovskite layer. Figure 3.12 illustrates the working of vapor assisted solution process. In this method, during the second step, MAI or FAI is vaporized to react with PbI_2. This method produces better contact between the precursors compared to traditional solution methods. Since there is no chance of partial dissolving of perovskite precursors, the stochiometry also be improved.

Chen et al. pointed out that the MAI vapor is applied to the previously prepared PbI_2 under an annealing temperature of 150 °C. They use glove box for their entire preparation. The film produced with this technique has large grain size, full phase formation, and improved film coverage. At the device level, it shows a PCE of 12.1%. The only drawback is its preparation time, which lasts for hours rather than minutes compared to spin-coating methods. This method is further modified by heating MACl powder at 100 °C, placed in a closed Petri dish that contains previously prepared $MAPbCl_{3-x}I_x$ on ITO/PEDOT: PSS substrate. This addition raised the PCE value to 15.1% and improved its stability up to 60 days. Vapor-assisted solution method can be considered one of the promising ways of fabricating perovskite films, which exhibit higher PCE at the device level. The major shortcoming of this method is the fabricating time, if we can reduce the time then this method can be considered for commercialization purposes [20].

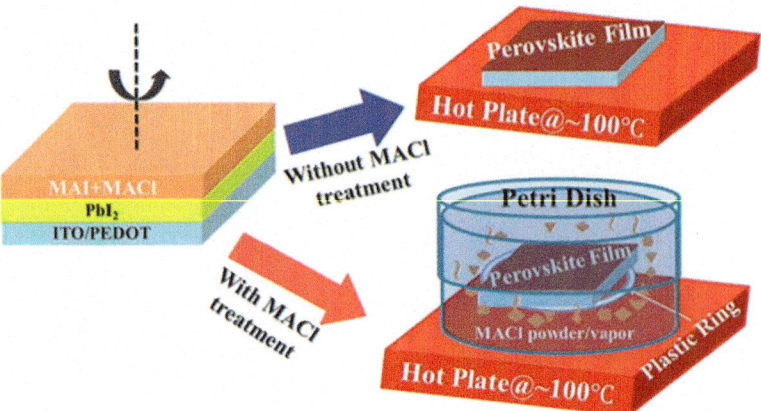

Fig. 3.12 Illustrates vapor assisted solution process (open access)

References

1. G. Zhao, H. Kozuka, H. Lin, M. Takahashi, T. Yoko, Thin Solid Films **340**, 125 (1999)
2. M.A. Green, A. Ho-Baillie, ACS Energy Lett. **2**, 822 (2017)
3. P. Wang, Y. Wu, B. Cai, Q. Ma, X. Zheng, W.H. Zhang, Adv. Funct. Mater. **29** (2019)
4. Deepika, A. Singh, U.K. Verma, A. Tonk, Physica Status Solidi (a) **220**, 2200736 (2023)
5. P. Arjun Suresh, G.S. John, A.M. Johnson, U.S. Sajeev, K.V. Arun Kumar, J. Mater. Sci.: Mater. Electron. **35** (2024)
6. J. Stenberg, Perovskite solar cells (2017)
7. G. Yang, H. Tao, P. Qin, W. Ke, G. Fang, J. Mater. Chem. A **4**, 3970 (2016)
8. K. Mahmood, S. Sarwar, M.T. Mehran, RSC Adv. **7**, 17044 (2017)
9. S. Li, Y.L. Cao, W.H. Li, Z.S. Bo, Rare Met. **40**, 2712 (2021)
10. Z. Shariatinia, Renew. Sustain. Energy Rev. **119** (2020)
11. S. Thomas, A. Thankappan, Perovskite photovoltaics (2018)
12. N.J. Jeon, J.H. Noh, Y.C. Kim, W.S. Yang, S. Ryu, S.I. Seok, Nat. Mater. **13**, 897 (2014)
13. J. Burschka, N. Pellet, S.J. Moon, R. Humphry-Baker, P. Gao, M.K. Nazeeruddin, M. Grätzel, Nature **499**, 316 (2013)
14. P. Kajal, K. Ghosh, S. Powar, Applications of solar energy (2018)
15. C.Y. Chang, Y.C. Huang, C.S. Tsao, W.F. Su, ACS Appl. Mater. Interfaces **8**, 26712 (2016)
16. S. Razza, S. Castro-Hermosa, A. Di Carlo, T.M. Brown, APL Mater **4** (2016)
17. J. Jiao, C. Yang, Z. Wang, C. Yan, C. Fang, Results Eng. **18**, 101158 (2023)
18. M. Liu, M.B. Johnston, H.J. Snaith, Nature **501**, 395 (2013)
19. Z. Yang, C.C. Chueh, F. Zuo, J.H. Kim, P.W. Liang, A.K.Y. Jen, Adv. Energy Mater. **5** (2015)
20. Z. Shi, A.H. Jayatissa, Materials **11** (2018)

Efficiency Enhancement of PSC

In the case of photovoltaic devices, the stability and efficiency are the crucial factors that determines the growth of solar cell. PSCs are deemed to be the latest innovation, with a notable increase in efficiency in short time. Now the efficiency of the device has reached nearly 26%, which is analogous to silicon solar cell existing in the market [1]. Because of its capabilities, PSCs are one of the solar cells that are most researched. The various layers used in the solar cell has distinctive characteristics that affect its stability as well as effectiveness. Apart from the efficiency figures stability is the most important part where PSCs lack its improvement. Several studies, theoretical and experimental works, are going on to improve this. According to current research, an assortment of factors, including oxygen, temperature, moisture, UV light, and others, may contribute to the destruction of PSCs [2].

4.1 Tolerance Factor

Among the different layers of PSCs, the most important layer which shows stability issues is the active layer. Thus increasing its stability enhances the overall working of solar cell. By figuring out the tolerance factor, one can get a clear picture of the extent of distortion in the perovskite structure. Determining the tolerance factor (t) of the PSCs is one of the best ways to predict the degradation rate theoretically. Finding the tolerance factor constant can measure the level of distortion that happens in the ideal cubic structure of the perovskite. It can vary from 0.75 to 1.1, values closer to 1 indicate the cubic structure.

Fig. 4.1 Perovskite structure (Open access)

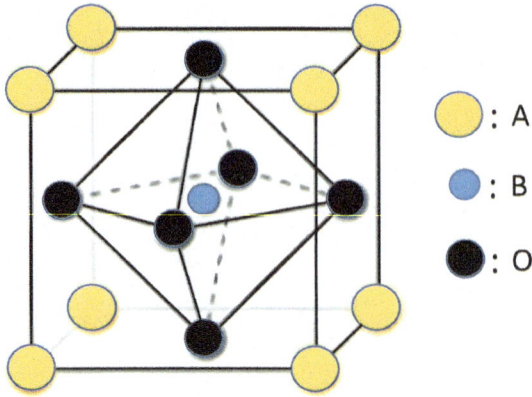

The lower tolerance factor of less than 1 indicates less symmetry in the crystal structure. A material's lattice constant and tolerance factor are useful in recognizing its symmetries and properties.

Figure 4.1 represents perovskite structure ABO_3, the A site is larger than the B site; the A site have an ionic radius of 1.2 Å, whereas the B site is 0.7 Å. In this 12 oxygen ions are coordinate with the A site and 6 oxygen ions are coordinated with the B site [3, 4].

Based on the Goldschmidt tolerance factor (t),

$$t = \frac{R_A + R_O}{\sqrt{2}(R_B + R_O)}$$

Where R_A, R_B, and R_O represent the ionic radii of the A-site cation, B-site cation, and oxygen anions, respectively. With a tolerance value nearby 0.9, $MAPbI_3$, one of the most extensively utilized perovskite materials, can crystallize in the black alpha phase. However, for $FAPbI_3$, the tolerance factor is 1, and at room temperature, it crystallizes into photoinactive delta phase [5].

4.2 Current Efficiency and Scope of Enhancement

The efficiency and reliability of solar cells are necessary for achieving success in the photovoltaic market. At the moment, Si solar cells have the highest efficiency of 25%. When PSCs are scaled up to larger devices, their efficiency is anticipated to surpass 30%. At the laboratory level the efficacy of PSCs has achieved about 31%, this is the crucial reason to see the perovskite solar cells by way of futuristic photovoltaic device [6].

4.3 HTL and ETL Dependence of Efficiency

In 2009, at the early ages of perovskite solar cell devices, Japanese scientists found that organic metal halides (MAPbI$_3$) are similar to dyes used in DSSCs and can absorb sunlight. When it is used as a sensitizer in DSSC along with liquid electrolyte the efficiency of the device becomes 3.8%. Later in 2012, Whenever solid electrolytes were used in place for liquid ones, solar cells' efficiency boosted to 9.7% (Spiro-MeOTAD). Hence it gained popularity, and more investigation is being done in this field. Additionally, this processes, certain advantages like high absorption rate, ease of fabrication, reduced recombination rate, high open circuit, and high internal quantum efficiency. In 2022, the efficiency was further improved to 20.1% and a fill factor of 84.3% was reported. This PSC is completely prepared using the spin-coating technique. The low-defect perovskite film showed a higher absorption rate, higher rate of photocarrier extraction and transport, which results in a higher fill factor. The maximum efficiency achieved by a multijunction perovskite solar cell is 30% for the $CH_3NH_3PbI_{3-x}Cl_x$ system. The Shocley-Queisser limit predicts a maximum theoretical efficiency of 31.4% for a 200 nm thick PSC.

4.3 HTL and ETL Dependence of Efficiency

Other than perovskite materials, the stability and effectiveness of PSCs depend on ETL and HTL. Upon light exposure, the active layer will create electron–hole pairs. The ETL and HTL layer which are close to the active layer conduct, electrons and holes corresponding to the electrodes. The inverted structure (p-i-n) and normal (n-i-p) structures have been introduced to improve the performance of PSCs. Common n-i-p organic HTL materials that are applied above the perovskite layer embrace porphyrins, PTAA, and Spiro-OMeTAD. Since light transmits through the HTL to get to the perovskite layer in the p-i-n structure, the transparency of the HTL is crucial [7].

The main functions of HTL are:

- Assist like an energy barrier between the perovskite layer and metal contact.
- Modifies the PSCs overall performance with the high hole mobility of HTL.
- Serves as a barrier against moisture and prevents degradation of the active layer [8].

The selection of HTL has a major effect on the PSC's stability. The productivity of a solar cell with an inverted structure that utilizes PEDOT:PSS (3,4-ethylenedioxythiophene-polystyrene) as the HTL, for instance, is over 18%. Nevertheless, the device stability is affected by the humidity and acidic nature of HTLs. When NiO_x is used in its place, the device preserves 90% of its efficiency after 60 days of air storage, with better stability and an efficiency of 16% [7].

4.3.1 Selection of HTL

The generation of current from photovoltaic devices involves four major procedures:

- Light absorption
- Creation of electron–hole pairs by active layer
- Separation of electrons and holes
- Transfer to corresponding electrodes.

Low recombination rate and rapid carrier transportation are the key factors for a superior solar cell. Carrying holes and hindering electrons from departing the active layer are the main tasks of HTL. Figure 4.2 shows the energy level diagrams of major inorganic materials.

When the perovskite layers absorb photons, the electrons in the material are excited into the LUMO and form charge carriers in the active layer. HTL or ETL can be utilized to convey the resulting free positive and negative charges to the appropriate electrodes [9]. The extraction of holes from the perovskite layer to the HTL and eventually to the ITO electrodes is seen in PSCs due to the HOMO level of the HTL is somewhat greater than that of the perovskite layer. Because of equivalent energy levels of HTL and perovskite, electron and hole depletion is prevented, and charge recombination losses are lowered.

4.3.1.1 Major Hole Transport Materials Used in PSC

An ideal HTL should possess good thermal stability, photochemical stability, high intrinsic hole mobility, and matching energy levels with the active layer. Regarding large-scale devices and for commercialisation the HTL should be economical and easy to prepare. In solid-state PSC, the HTL was first introduced in 2012. Spiro-OMeTAD is a popular HTL that chosen today for high-performing solar cells. Spiro-OMeTAD's disadvantage is that

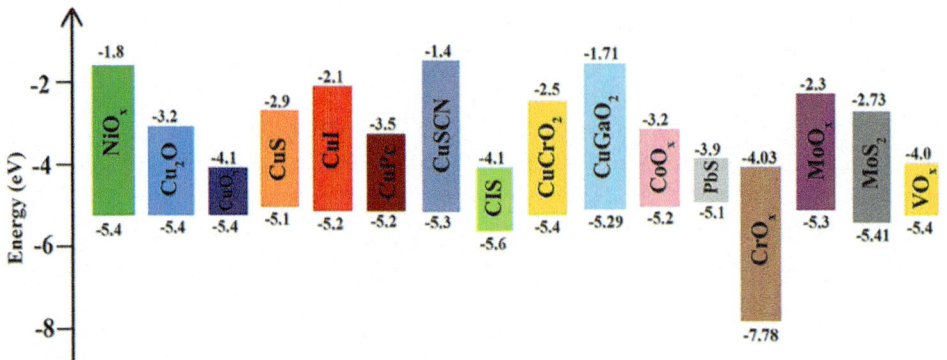

Fig. 4.2 Energy level diagram of inorganic HTLs (Copyright)

4.3 HTL and ETL Dependence of Efficiency

it can only be used with particular additives, such as tBP (4-tert-butylpyridine) and Li-TFSI (bis(trifluoromethane) sulfonimide lithium salt) [10]. Additionally, certain dopent-free organic hole transport materials (HTM) have been developed, such as PEDOT: PSS and P3HT. The major disadvantage of organic HTMs is their lack of stability compared to inorganic HTMs.

Spiro-OMeTAD

Spiro-OMeTAAD was first used as a heterojunction layer in DSSC resulting in high efficiency. Because of its greater efficiency values, PSCs also gained a lot of prominence. The device that uses spiro-OMeTAD as HTL exhibits an efficiency of more than 22%. As a downside of the material, the devices exhibit very little stability. Owing to its amorphous nature and its unique chemical structure, it shows instability in factors such as water, light and heat. To address this issue, various dopants were used to enhance its stability factor. Several studies were conducted by doping fluorine on spiro-OMeTAD to enhance its stability. The researchers studied the influence of structural modifications of 4,4,'-dimethoxydiphenylamine-HT1, 2,7-dibromo-9H-fluorene, N-bis(4-methoxyphenyl)aniline, catalysed by Pd(dba)2-HT2. The higher molecular weight and extended three-dimensional structure of HT2 helps to increase the photovoltaic performances but not stability. It is found that doping with copper salts like CuI or CuSCN in Spiro-OMeTAD enhanced its stability up to some extent. The conductivity and hole mobility also increased with doping. This results in an efficiency improvement from 14.82 to 18.02%. Adding copper salts also reduced the aggregation and crystallization of spiro-OMeTAD films which results in lesser pinholes and voids. As a result, the decomposition rate of perovskite decreases to some extent [11].

Utilizing polymers for HTL

PEDOT: PSS

An HTL that is commonly used in PSCs are PEDOT: PSS. It has several advantages which include its low hysteresis and ease of preparation. Its acidic nature and the high energy barrier between the HTL and perovskite layer, nevertheless, cause very poor photovoltaic performance. Due to these disadvantages, it shows relatively low V_{OC} and short circuit current J_{SC}. The device prepared with PEDOT: PSS achieved up to 18.1% efficiency in the IPSC configuration (ITO/PEDOT: PSS/MAPbI3/PCBM/Au) in 2015. A further investigation on doping polymer electrolyte (PSS-Na) into PEDOT: PSS was published in 2017. While compared to undoped, the doped material depicts a higher work function and greater energy level matching [10]. PCE grows from 12.35 to 15.56%, and V_{OC} is modified from 0.96 to 1.11 V. In accordance with another research in 2018, doping PEDOT: PSS with graphene oxide (GO) improves the PCE to 18.09% while additionally demonstrating a notable improvement in stability. Later, a facile solvent engineering approach was reported, which takes out the predominant PSS from PEDOT: PSS, resulting in a PCE of 18.18%. This approach makes the film non-wettable, more stable, exhibits higher carrier lifetime, and shorter charge transport time compared to others.

P3HT

Another polymer HTM utilized in PSCs is P3HT (Poly(3-hexylthiophene). It is promising due to its low cost, high charge carrier mobility, and compatible bandgap. But, owing to its low conductivity, it reveals a low PCE value at the device level. Several studies were conducted to increase the conductivity of the material. In 2016, an effective P-type dopant, F4TCNQ was introduced into P3HT, resulting in an increase in conductivity on the bulk level. Compared to undoped P3HT, which has a PCE of 10.3%, the doped P3HT used in mesoscopic PSCs shows a PCE of 14.4% [10]. It is found that doping 1% of F4TCNQ increases the conductivity up to 50 times. F4TCNQ doped device also shows higher stability at a humidity of 40%. Also reported that there is a relation between the photovoltaic characteristics and molecular weight (MW) of P3HT. When the molecular weight of P3HT increases the efficiency of the PSCs also increases. The improvement in fill factor (FF) and J_{SC} is additionally linked with the increased efficiency. When MW is increased, it enhances the electron lifetime and absorbance, thereby increasing the overall efficiency at the device level [11].

PTAA

Another polymer HTM that is often used in PSCs is poly[bis(4-phenyl)(2,4,6-trimethylphenyl)amine]. Since it has some stability issues it works only with certain additives, like in the case of Spiro-OMeTAD. Doping PTAA with Li-TFSI and tBP shows the highest efficiency of 22.1% and 19.7% respectively in the case of 1 cm^2 cell [12]. To solve the stability issue of PTAA, it was doped with LAD (Lewis acid dopant) instead of Li-TFSI and tBP, resulting in an efficiency of 19.01%, compared to 17.77% for Li-TFSI and tBP doped device. The doping also reduced the J-V hysteresis and increased the overall stability. Inverted PSCs (p-i-n) additionally incorporate PCBM ([6,6]-phenyl-C61-butyric acid methyl ester) as ETM and PTAA as HTL. These devices show a higher lifetime and no hysteresis, although the PCE of the device is lesser relative to the n-i-p configuration [10].

Inorganic Hole Transport Materials

In PSCs, organic materials exhibit excellent PCE but it lacks stability. To address this issue inorganic HTM can be used, inorganic HTM has various advantages including its higher stability, lower cost and simple sample preparation. One of the downsides of inorganic HTMs is their lower PCE values. There don't seem lots of research on inorganic materials that report PCEs greater than 20% [10].

4.3 HTL and ETL Dependence of Efficiency

Cu_2O as HTM

Cu_2O (cuprous oxide) and CuO (cupric oxide) are the mostly used inorganic HTMs. In inverted PSCs, it is frequently chosen over PEDOT: PSS for the reason of its higher absorption coefficient. It was reported in 2016 that an efficiency of 11% had been achieved via the use of Cu_2O as an ultrathin film in planar inverted PSC. The Cu_2O prepared has good energy level alignment with the active layer $MAPbI_3$ ($CH_3NH_3PbI_3$). Later CuO_X layer was produced with a facile route of the solution-processed method. The film thus obtained maintains 90% of transparency and has higher stability. Using this layer in PSC yielded a PCE of 17.1%. Also investigated the variation of performance in the ratio of Cu^{2+} and Cu^+ in CuO_X and found that the PCE of the device remained the same while the fill factor (FF) decreased with CuO content. Cu/Cu_2O film was generated by ion sputtering and subsequently incorporated between the Ag electrode and spiro-OMeTAD in mesoporous PSCs, yielding a PCE of 17.11%, according to a different study out in 2018. Introducing the Cu_2O layer, also increases the stability of the solar cell by acting as a barrier against moisture [10].

CuSCN as HTM

CuSCN, or copper(I) thiocyanate, has the benefit of exceptional transparency in the visible spectrum in addition to its high chemical stability and ease of preparation. Notwithstanding the benefits, the solvent DES (diethyl sulfide), which is typically used to process CuSCN, decays the perovskite layer, hindering its potential for employing in n-i-p-based CuSCN PSCs. Therefore, scientists attempted to build PSCs based on p-i-n CuSCN. The CuSCN layer was created by electrodeposition in previous studies, and then a $CH_3NH_3PbI_3$ film was deposited on it by employing the fast deposition-crystallization method. The resulting perovskite layer had low surface roughness and low interface contact resistance [10]. Through an optimal PCE of 16.6%, the CuSCN-based PSCs that were thereby created had an average PCE of 15.6%. This improvement in performance can be attributed to CuSCN HTL's high hole mobility.

Furthermore, some methods have been indicated to successfully mitigate the perovskite layer damage that appears when CuSCN is processed for n-i-p PSCs. The PCE has gone up by more than 20% as a result. Jung and associates generated a low-temperature solution-processed CuSCN HTM in 2016 and used on mesoporous PSCs. The resulting CuSCN proved a PCE of 15.9% with a peak yield of 18%, and its crystalline structure remained very stable even at high temperatures. In addition to having outstanding thermal stability, after two hours of annealing at 125 °C in air with 40% humidity, roughly 60% of the initial PCE remains present [10]. Conversely, under comparable circumstances, spiro-OMeTAD-based devices were only capable to maintain 25% of the PCE.

By using quick solvent removal techniques, Arora et al. later lessened the harm that DES caused to the perovskite layer. CuSCN layers have been effectively manufactured to be compact and highly conformal, allowing for the rapid extraction and gathering of charge carriers. With a PCE of more than 20%, the mesoporous PSCs employing CuSCN as HTL suggested excellent thermal stability [10]. The possibility of induced degradation of the CuSCN/Au contact limits its operational stability, so a graphene oxide spacer layer was installed between CuSCN and Au [13]. As a consequence, devices with rGO and CuSCN obtained a 20% PCE [10]. This was nearly identical to the spiro-OMeTAD devices that were previously discussed. Under both thermal stress and continuous full-sun illumination, the device's stability was significantly higher than that of spiro-OMeTAD devices, according to studies on its structure and photovoltaic performance. The MAPbI$_3$ layer was surfaced with various functional molecules by Young et al. in 2019 in order to eliminate the defects while enhancing the contact between the MAPbI$_3$ and CuSCN layers. As a result, CuSCN-based PSCs' V$_{OC}$ improved.

The V$_{OC}$ of CuSCN-based PSCs raised to over 40 mV with the addition of phenylene-1,4-di-isothiocyanate (Ph-DITC) and 3-pyridyl isothiocyanate (Pr-ITC), producing an average PCE of 18.5% and a peak PCE of 19.17%. Devices with this interfacial modification displayed more effective long-term stability and less J-V hysteresis than pristine devices. Kim et al. used polydimethylsiloxane (PDMS) as a polymeric interlayer in CuSCN-based perovskite solar cells to enhance conversion efficiency and stability and avoid interfacial degradation. They discovered that PDMS could form chemical bonds with CuSCN and perovskite, which improved hole transport at the interface and passivated the interfacial defects [10]. In addition to having a higher PCE of more than 19%, the PSCs with the PDMS layer displayed improved durability against temperature and humidity.

Copper (I) iodide
Another popular inorganic HTM is CuI, which has a substantial bandgap, strong chemical stability, low cost of production, and high hole mobility. In 2014, Christain et al. used CuI in mesoporous PSCs for the first time and they gained a promising PCE of 6%. The fill factor (FF) of CuI-based devices was higher compared to spiro-OMeTAD-based devices, but the V$_{OC}$ was lower [10]. This is because CuI has a higher conductivity than spiro-OMeTAD. In contrast to spiro-OMeTAD, CuI had an electrical conductivity that was two orders of magnitude higher. Using CuI as the HTM, Sepalage et al. offered a planar n-i-p PSC in 2015. These devices showed no discernible hysteresis, with a maximum PCE of 7.5% and an average PCE of 5.8 [14]. Chen et al. used solution-processed PEDOT: PSS or CuI as HTM to create inexpensive inverted [p-i-n] planar PSCs. Nevertheless, there was no discernible improvement in PCE. The obtained value was 13.28% for PEDOT:PSS-based devices and 13.58% for CuI-based devices. However, substituting CuI for PEDOT:PSS significantly improved the device's stability. Li et al. created mesoporous PSCs in 2017 by employing CuI film as the HTL and modified TiO$_2$ as the ETL [10]. In order to improve ETL's electron conductivity and mobility, they modified the TiO$_2$ layer using Na. Here, they implemented the easy spray deposition method to synthesis the CuI layer [15]. With exceptional stability and minimal hysteresis, the device resulted in a maximum PCE of up to 17.6%. In 2019 Cao et al. created a hybrid CuI/Cu nanostructure using the partial

4.3 HTL and ETL Dependence of Efficiency

iodation of Cu nanowires method, and it was then employed as HTM in the inverted [p-i-n] PSCs. While the inner Cu promoted the quick transfer to the extracted charges, the outer CuI assisted with the beneficial charge extraction. Furthermore, by employing an amalgam of PCBM and ZnO nanowires as ETL, they could increase the devices' stability. The obtained devices produced a maximum PCE of up to 18.8% and shown excellent long-term stability against humidity [10].

Nickel oxide (NiO_x)

NiO_x exhibits high chemical stability and has a deep valance band. NiO_x initially served as HTM in traditional (n-i-p) PSCs. It has gained a high PCE of over 20% thus far. The study by Wang et al. (2014) looked into the way oxygen doping affected low-temperature sputtered NiO_X films and how it impacted device performance. According to their research, Ni^{3+} is created throughout the process. They discovered that while a higher amount of Ni^{3+} will impair performance, the proper amount can enhance the PCE of PSC devices. With appropriate doping at a 10% oxygen flow ratio, they gained a PCE of 11.6%. Xu et al. (2015) reported PSCs with a carbon electrode, a mesoporous ZrO_2 layer acting as a space separator, and mesoporous NiO and TiO_2 layers acting as hole transport and electron transport contacts, respectively. Ultimately, a 14.9% PCE was attained. Cao et al. created NiO-based PSCs with a similar structure and used Al_2O_3 as a space separator to achieve a PCE of 15.03%. Additional studies on improving the photovoltaic performance of NiO-based PSCs revealed that inverted (p-i-n) devices outperformed conventional (n-i-p) ones [10]. Xie et al. published their studies based on the FTO/NiO_X/$FAPbI_3$/PCBM/TiO_X/Ag PSC configuration. Through adding methylammonium chloride (MACl), a high crystallinity FA-based perovskite film was created in 2017. The perovskite's vertical recrystallization was made easier by the addition of MACl, which gave rise to greater film quality. As a result, the PCE boosted to 20.65% and the carrier recombination of the perovskite layer decreased. Also, the devices showed excellent thermal stability and light soaking stability owing to the low methylammonium (MA) content, phase-pure morphology, and highly crystalline structure in perovskite films. Yue et al. introduced several schemes for efficiently controlling the charge extraction in inverted NiO_x-based PSCs. The V_{OC} of the devices raised with doping Cl^- into the perovskite layer of $CH_3NH_3PbI_{3-x}Cl_x$. The bandgap alignment of the devices was enhanced by amending the aluminium (Al) cathode with zirconium acetylacetonate, doping copper into the NiO_X HTL, and implementing an advanced FTO substrate [10]. The extraction of charge carriers was greatly enhanced by these tactics, which efficiently reduced trap states and raised the transport rate. Accordingly, a remarkable 20.5% power conversion efficiency (PCE) was gained.

4.4 Graphene as Conducting Electrode, HTL and ETL

Graphene is a 2-D arrangement of carbon atoms arranged in a honeycomb pattern (as shown in Fig. 4.3). rGO and GO are the most studied carbon materials and compared to pristine graphene it exhibits different properties.

Pristine graphene is known for its conductivity, it shows an electrical conductivity of 10^6 S cm^{-1}. While rGO is conductive by nature, it is not as conductive as pristine graphene. GO, on the other hand, is believed to be an insulating substance with extremely little conductivity. These features make it an attractive option for conducting electrodes, HTMs and ETMs in solar cells.

4.4.1 Conducting Electrodes

In the case of PSC, the conducting electrode plays a major part in cell efficiency and stability. Generally Au, Ag and Al are mostly used which is deposited through vapour deposition methods. Carbon-based materials have gained significant attention for developing conducting electrodes of high efficiency, stability and low cost. In a recent study, carbon nanotube (CNT) was used as the top electrode of PSC. Since the sheet resistance of CNT ranges from 2 to 25 kΩ sq^{-1}, which is relatively high compared to other electrodes, it negatively affects the device performance. Even though it has some disadvantageous, it offers higher long-term stability. To increase the conductivity, the CNT layer had to be much thicker, which will reduce the transparency of the device. An efficiency of 15.36% is reported for PSC which uses graphene/CNT as counter electrodes. It maintains an efficiency of 86% of initial PCE when it is stored in ambient conditions for 500 h. The hybrid graphene/CNT system offers excellent conductivity and moisture-blocking capability compared to pristine CNT and Ag systems [16].

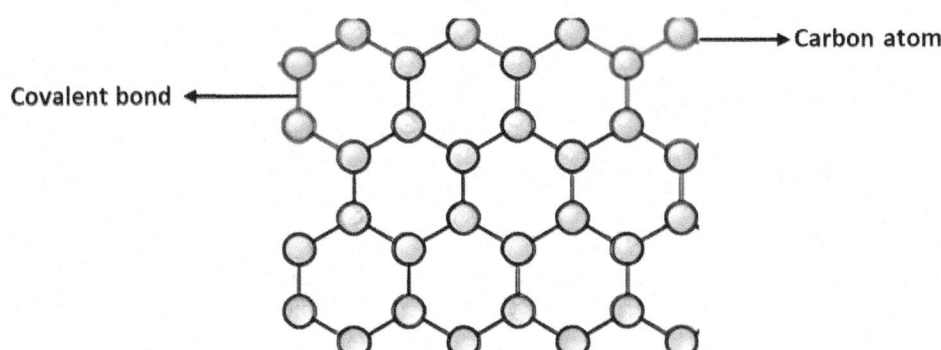

Fig. 4.3 Structure of graphene

4.4 Graphene as Conducting Electrode, HTL and ETL

Graphene is also used as a conducting electrode in PSCs. Due to the high resistance of single-layered graphene (~ 1050 Ω sq^{-1}), a single layer of PEDOT: PSS doped with fluorosurfactant zonyl-FS100 and D-sorbitol is spin-coated (thickness 20 nm) over the graphene layer [17]. Implementing PEDOT: PSS to graphene will: (i) decrease the material's sheet resistance; (ii) serve as an adhesive during the lamination process; and (iii) increase the graphene's hole doping. The device earned an efficiency of 12.37% while using the double-layered graphene as conducting electrode. In another investigation, used single-layered graphene as a transparent bottom anode in a device of structure, substrate/PEDOT: PSS/MAPbI$_3$/C$_{60}$/BCP/LiF/Al [10]. But due to the highly hydrophobic nature of graphene, neither perovskite nor PEDOT: PSS coats above the graphene layer. Incorporating a layer of MoO$_3$ (molybdenum trioxide) of 2 nm thickness on the graphene layer has achieved an efficiency of 17.1%. The increase in efficiency is due to the use of MoO$_3$ as HTM which increases the hydrophilicity of the graphene layer and also leads to desired energy level matching between MoO$_3$-graphene and PEDOT: PSS [17]. This structure was also tried on a flexible PEN (polyethylene naphthalate) substrate and observed its operational stability by contrast with repeated bending. After 1000 bending cycles with a radius of 6,4,2 mm, the device exhibits a PCE of 16.8% for the configuration PEN/graphene/MoO$_3$/PEDOT: PSS/MAPbI$_3$/C$_{60}$/BCP/LiF/Al, which is nearly 90% of its initial efficiency [18]. The graphene and perovskite layers appear to be undamaged based on the test results. At a bending radius R = 4 mm, however, the device efficiency for the ITO flexible substrate dropped drastically after 250 bending cycles [19]. Repeated bending causes cracks on the thin ITO layer and it spreads onto the perovskite layer. Relative to graphene substrate, the PEN/ITO substrates offer a sheet resistance of 13.3 Ω sq^{-1}, which is noticeably lesser than 552 Ω sq^{-1} of MoO$_3$/PEN substrate. This leads to low charge collection efficiency, low fill factor (FF), low series and high shunt resistance. However, both devices display comparable J$_{SC}$ results, which may be due to the higher transparency (~ 97% in visible range) offered by graphene-based substrate compared to ITO substrate (~ 89%).

In an alternative investigation, the conducting electrode is a graphene-coated PET (polyethylene terephthalate) substrate. The substrate is 20 μm thick, and the HTL is made of P3HT. The reason to choose P3HT as HTL is that it has a HOMO level of ~ − 5.2 eV, which is close to the valance band of MAPbI$_3$ (~ − 5.4 eV) [17]. However, the HOMO level of the most widely used HTMs, like PEDOT: PSS, is about 5.0 eV. Since P3HT is hydrophobic, the device's stability was also improved, and its efficiency was 10.4%. But when a ~ 2 μm layer of polymer ZEOCOAT^{TM} (cross-linkable olefin type polymer) is introduced before coating graphene, enhanced its efficiency to 11.5%. The increase in efficiency is due to the reduced roughness when the polymer is integrated. Aside from that, the device's performance deteriorated by only 14% after 500 bending cycles at a radius of 0.175 cm in the course of bending cycles.

4.4.2 Materials Employed in Hole Transport

Additionally, hole transport layers are modified using graphene and its derivatives. High hole mobility, well-aligned energy gaps, and high thermal stability are essential features of an ideal HTL. The most extensively used HTLs are Spiro-OMeTAD and PEDOT: PSS, but given their hygroscopic nature, they have difficulties with stability. Incorporating CNT which are highly hydrophobic can address this stability issue up to some extent. Researchers used poly(methylmethacrylate) (PMMA) covered with CNT sheet as HTL and the device exhibited a 5.82% of efficiency rate. PMMA shrinkage, which boosts the contact area and minimizes the resistance within perovskite and CNT, is responsible for of the lower PCE. With $V_{OC} = 1.45$ V, the device outperformed the widely used Spiro-OMeTAD. Moreover, the CNT rarely shows charge carrier selectivity. This issue can be solved by blending HTL materials into CNT films. An HTL is incorporated into an n-i-p device by coating SWCNT (Single-walled carbon nanotubes) with P3HT and PMMA. The p-type characteristics of P3HT enhanced the electrical properties of SWCNT. Additionally, PMMA is used as an encapsulant, applied above the P3HT/SWCNT layer, and it effectively blocks humidity and oxygen from entering the perovskite layer. The efficiency of this device reached up to 14.2% with excellent stability [20].

The most commonly used HTL in regular planar PSCs is Spiro-OMeTAD; in the present research, carbon derivatives were blended into the material to improve its stability and efficiency. In general, to increase the stability and efficiency of Spiro-OMeTAD certain hydrophilic additives are used. In this study, graphene-CNT composite mixed with Spiro-OMeTAD was used in planar PSC. Spin coating has been employed to introduce the CNT-Graphene composites to spiro-OMeTAD. Over the MAPbI$_3$ perovskite layer, the mixed composites had been spin-coated. It shows higher PCE and J_{SC} compared to pristine CNT [21]. In another study to reduce the recombination in the perovskite-HTL interphase, rGO was implemented between perovskite and HTL spiro-OMeTAD. This method is called the passivation surface method and it enhances hole extraction kinetics. The device prepared by this method gave an efficiency of 18.75% [22].

CuSCN is another material that is commonly used as HTL. It offers higher thermal and moisture stability and high hole mobility at a lower cost. However, the permeable properties of CuSCN reduce device stability due to ion migration and moisture diffusion. A PCE of 14.28% was established in recent studies on inverted PSCs utilizing a bilayer HTL made of rGO and CuSCN. Even after 100 h of exposure to AM 1.5 sunlight, it was discovered that the addition of rGO improved device stability and preserved 90% of its initial efficiency. Incorporating a rGO spacer layer between the CuSCN and the Au electrode in the mesoscopic architecture also results in an increase in stability [22]. Even after 1000 h of sun exposure at an ambient temperature of 60 °C, the CuSCN/rGO/Au interface device maintains 95% of its initial efficiency. However, the device that did not use rGO as a spacer layer lost half of its initial efficiency. The undesired interaction

4.4 Graphene as Conducting Electrode, HTL and ETL

between Au electrode and thiocyanate anions leads to device degradation. Another study reported the influence of rGO integration with CuSCN in inverted planar solar cells.

With an ideal CuSCN layer thickness of 10 nm, it became apparent that the series resistance of 9.7 Ω for the rGO single HTL was decreased to 4.9 Ω in the case of the bilayer HTL [22]. The rGO/CuSCN bilayer HTL provides a better interaction with the perovskite active layers versus the rGO single HTL, leading to larger grain size and improved PL intensity quenching. The device's stability improved when CuSCN and Au were separated by an interfacial layer of CVD-grown graphene. The further addition of graphene will work as a barrier to prevent moisture, I-ion migration, and Au electrode diffusion. When stored in a dark environment with 50% relative humidity, the CuSCN/G/Au interface device display 94% of its initial efficiency. In the perovskite/HTL interface, graphene serves as an efficient conductive barrier to holes [22].

4.4.3 Material for Transporting Electrons

ETLs' primary function is to block holes created by the active layer and conduct electrons. When identifying an electron transport material, the biggest need is that the material's conduction band minimum (CBM) be lower than that of perovskites. One of the ETLs most frequently found in DSSCs and organic solar cells (OSCs) is TiO_2. TiO_2 has instability despite its ability to block holes effectively. When exposed to the full spectrum of radiation, it will begin to degrade. Its efficiency and stability are limited by photogenerated holes created when it is exposed to UV light. These holes react with adsorbed oxygen at surface vacancies and operate as trap sites [22]. By adding graphene, which can be created by liquid-phase exfoliating graphene flakes, to mesoporous TiO_2, the device's long-term stability under extended light exposure is greatly improved. At the TiO_2 + G/perovskite interface, graphene's adsorption onto TiO_2 creates favourable interactions. The device retains over 88% of device efficiency after 16 h. Graphene in contact with perovskite may result in interfacial ferroelectricity and may increase the efficiency of electron collection [22]. Additionally, this process shrinks electron–hole recombination and shifts the hole wave function from pristine graphene to the TiO_2 perovskite interface. Another study found that using graphene flakes doped with TiO_2 as ETL raised more electron collection from the perovskite layer. G-doped TiO_2 also shows a lower decay time constant of 15 ns as compared to 25 ns in undoped TiO_2. This result was found from analysing photoluminescence intensity (PL), and the results show increased electron collection as compared to undoped.

If TiO_2 has low electron mobility, charge accumulation occurs at the TiO_2-perovskite interface and causes degradation due to the reaction between the perovskite layer and TiO_2. Alternatively, doping TiO_2 with graphene reduces charge accumulation by improving charge injection and transport at the interface. The corresponding reduction in charge accumulation will improve the device's overall stability and efficiency while slowing down the rate of degradation. However, compared to a pristine TiO$_2$ device, faster degradation can be seen in a mTiO$_2$ + G ETL device after losing 15% of initial efficiency when heated at 60 °C in the oven. These results show that the thermal stress caused by prolonged exposure to heat can permanently degrade the interaction between TiO$_2$ and the perovskite layer due to the burn-in effect [22].

A ZnO is taken into consideration as a substitute for the TiO$_2$ electron transport layer. It has several advantages, including high UV exposure stability and the ability to be prepared at lower temperatures, which makes it appropriate for flexible solar cells. However, ZnO has lower PCE values than TiO$_2$. When it interacts with the perovskite layer, it exhibits chemical fragility and gradually changes the perovskite crystal structure into lead halide. The deprotonation of MAI with the hydroxyl group on the ZnO surface is what causes this process [22]. ZnO's interfacial defect will lead to a higher recombination rate and reduce the device's performance. Doping with suitable materials is an alternative method to reduce the defects and to improve the electron collection efficiency. Graphene is a suitable material for doping, Chandrasekhar et al. incorporated different concentrations of graphene into ZnO and it was then coated as ETL in MAPbI$_3$ solar cell via spray deposition method. They achieved a PCE of 10.34% for 0.75 wt% of G-ZnO nanocomposites. Later, by adding nitrogen-doped graphene via a hydrothermal process, they enhanced the integration of graphene in the ZnO nanorod electron transport layer as it was prepared. The ideal concentration for the NG-ZnO nanorod nanocomposites was 0.8 wt%, and this modification produced a power conversion efficiency of 16.82%. The grain size of perovskite crystals increased in accordance to increase in graphene concentration adsorbed in ZnO. As a result, graphene provides a vital part in perovskite film growth. A larger grain size for the perovskite layer is essential to improving the charge collection efficiency [22]. Considering both studies, the optimal concentrations for graphene-doped ZnO were found to be 0.75% and 0.8%, achieving better PCE values. Nevertheless, photovoltaic performance declined when the concentration was raised above these thresholds. This decrease is explained by the buildup of graphene at the ETL/perovskite interface, that causes in the excess graphene coming into direct contact with the perovskite film. This results in lower electron collection efficiency and a minor drop in short-circuit current density (J_{SC}) [22].

Another substance that can be utilized in PSCs as an electron transport layer is SnO$_2$. Its broader bandgap of 3.6 to 4 eV gives it several advantages over TiO$_2$ and ZnO, as well as low trap density and bulk electron mobility. This wide bandgap results in higher stability under illumination. But when FTO is used as a substrate, the stability of the device decreases due to the transfer of fluorine to the SnO$_2$. It will reduce the electron selectivity of the SnO$_2$ layer and will decrease the overall efficiency of the solar cell. These restrictions limit its application to flexible and planar PSCs. However, due to the high annealing temperature of 150 to 200°C of SnO$_2$, it is challenging to use in PET and PEI(polyetherimide) substrates. Because of oxygen vacancies, SnO$_2$ prepared at low temperatures typically exhibits some trap states, which could cause hysteresis in the photovoltaic performance of the device. Zhu et al. attained a PCE of 18% with attenuated hysteresis in planar PSCs by doping SnO_2 with mechanically exfoliated graphene from graphene flakes. The hysteresis calculation shows only 0.2 for graphene-doped and 0.8 for pristine SnO_2 devices. The charge transfer between the SnO_2 and perovskite interface is significantly higher in the graphene-doped compared to undoped. Apart from that, adding graphene to SnO_2 improves device stability; for 300 h at 40% relative humidity, the device maintains 90% of its initial efficiency without encapsulation. In a different study, graphene nanosheets were doped into SnO_2 and used as the ETL to create a flexible PSC. With very little hysteresis, the device exhibits an efficiency of 13.36%. Low crystallinity is frequently the result of the SnO_2 being annealed at low temperatures during the creation of flexible PSCs. This will reduce the efficiency and raise the possibility of charge recombination at the ETL/perovskite interface. However, the incorporation of graphene into the SnO_2 increases the electron mobility in the ETL and ensures faster electron injection [22].

4.5 Influence of Graphene in Perovskite Materials

In PSCs, perovskite material is utilized as an active layer. Other than its ambipolar properties to conduct holes and electrons simultaneously, the film quality is also essential for better device performance. The morphology of the layer, thickness and grain size are some of the factors that influence the performance. Higher thickness and larger grain size of perovskite material can harvest maximum amount of light. Doping graphene materials can control the grain size of the perovskite and thereby increase the efficiency. Li et al. discovered that adding graphene nanowires to the perovskite material $MAPbI_3$ increases the nucleation and crystallization of the nanofibers. The PCE increased to 19.83%, and the grain size doubled in size, as stated in the results. In the case of stability, a 300 h exposure in ambient conditions reduced the initial efficiency by 10.5%. While when GO is embedded with the perovskite material in optimum concentration, it improves charge separation as it serves as hole accepter and increases the efficiency of the device to 15.2%.

N-rGO (nitrogen-doped reduced graphene oxide) had been introduced to the perovskite material $FA_{0.85}MA_{0.15}Pb(I_{0.85}Br_{0.15})_3$ in another study by Hadadian et al. The doped nitrogen will interact with the FA cations to produce a thicker perovskite layer and larger grain size, which improves the PSC's capacity to harvest light. By enhancing hole selection and minimizing the formation of charges in the perovskite layer, the N-rGO raises the device's open circuit voltage (V_{OC}). Kim et al. also found that introducing pristine rGO into the perovskite material also increases the V_{OC} values. Later Fang et al. increased the PCE values to 17.62% for the mesoscopic PSC by incorporating graphene quantum dot (GQD) into the perovskite material. The GQD introduced will passivate the electron trap and increases the efficiency. Also, the perovskite material where GQD is introduced shows lower charge transfer resistance (R_{ct}), this will increase efficient electron transfer between the perovskite/ETL interfacial layers.

References

1. F. Cao, L. Bian, L. Li, Energy Mater. Devices **2**, 9370018 (2024)
2. S. Nair, S.B. Patel, J.V. Gohel, Mater. Today Energy **17** (2020)
3. N. Ali, N. Shehzad, S. Uddin, R. Ahmed, M. Jabeen, A. Kalam, A.G. Al-Sehemi, H. Alrobei, M.B. Kanoun, A. Khesro, S. Goumri-Said, Int. J. Energy Res. **45**, 19729 (2021)
4. M.L. Rojas-Cervantes, E. Castillejos, Catalysts **9** (2019)
5. M.K. Rao, D.N. Sangeetha, M. Selvakumar, Y.N. Sudhakar, M.G. Mahesha, Sol. Energy **218**, 469 (2021)
6. S. Thomas, A. Thankappan, Perovskite photovoltaics (2018)
7. P.K. Kung, M.H. Li, P.Y. Lin, Y.H. Chiang, C.R. Chan, T.F. Guo, P. Chen, Adv. Mater. Interfaces **5** (2018)
8. M.L. Petrus, A. Music, A.C. Closs, J.C. Bijleveld, M.T. Sirtl, Y. Hu, T.J. Dingemans, T. Bein, P. Docampo, J. Mater. Chem. A **5**, 25200 (2017)
9. G.M. Arumugam, S.K. Karunakaran, C. Liu, C. Zhang, F. Guo, S. Wu, Y. Mai, Nano Select **2**, 1081 (2021)
10. S. Li, Y.-L. Cao, W.-H. Li, Z.-S. Bo, Rare Met. **40**, 2712 (2021)
11. S. Pitchaiya, M. Natarajan, A. Santhanam, V. Asokan, A. Yuvapragasam, V. Madurai Ramakrishnan, S.E. Palanisamy, S. Sundaram, D. Velauthapillai, Arab. J. Chem. **13**, 2526 (2020)
12. A. Rajagopal, K. Yao, A.K.-Y. Jen, Adv. Mater. **30**, 1800455 (2018)
13. V. Manjunath, S. Bimli, P.A. Shaikh, S.B. Ogale, R.S. Devan, J. Mater. Chem. C **10**, 15725 (2022)
14. G.A. Sepalage, S. Meyer, A. Pascoe, A.D. Scully, F. Huang, U. Bach, Y.-B. Cheng, L. Spiccia, Adv. Funct. Mater. **25**, 5650 (2015)
15. J.-Y. Shao, D. Li, J. Shi, C. Ma, Y. Wang, X. Liu, X. Jiang, M. Hao, L. Zhang, C. Liu, Y. Jiang, Z. Wang, Y.-W. Zhong, S.F. Liu, Y. Mai, Y. Liu, Y. Zhao, Z. Ning, L. Wang, B. Xu, L. Meng, Z. Bian, Z. Ge, X. Zhan, J. You, Y. Li, Q. Meng, Sci. China Chem. **66**, 10 (2023)
16. E. Muchuweni, B.S. Martincigh, V.O. Nyamori, Adv. Energy Sustain. Res. **2** (2021)
17. E.L. Lim, C.C. Yap, M.H.H. Jumali, M.A.M. Teridi, C.H. Teh, Nanomicro Lett. **10**, 27 (2017)
18. K. Petridis, G. Kakavelakis, M.M. Stylianakis, E. Kymakis, Chem. Asian J. **13**, 240 (2018)
19. M. Hadadian, J.-H. Smått, J.-P. Correa-Baena, Energy Environ. Sci. **13**, 1377 (2020)
20. U. Asghar, M.A. Qamar, O. Hakami, S.K. Ali, M. Imran, A. Farhan, H. Parveen, M. Sharma, Micromachines (Basel) **15** (2024)
21. J. Zhang, J. Fan, B. Cheng, J. Yu, W. Ho, Solar RRL **4** (2020)
22. N.E. Safie, M.A. Azam, M.F.A. Aziz, M. Ismail, Int. J. Energy Res. **45**, 1347 (2021)

Challenges of Preparing PSC 5

The PCE of Perovskite solar cells (PSC) is increasing rapidly. Apart from their high efficiency, they face several issues that prevent their commercialization. The major problems in commercialisation are their instability, reproducibility, the environmental impacts of lead-based PSCs, hysteresis, and the difficulty of preparing large-scale devices. Small changes in morphology, pinholes, and film roughness significantly influence the overall performance of PSCs [1].

5.1 Difficulties in Preparing PSC

The dependence of the performance of PSCs on various factors mentioned above makes its preparation more challenging. In the initial days of PSC research, the focus was mainly on increasing the surface coverage of perovskite films by varying the stoichiometry ratio of precursors. Thereby, taking a ratio of 3:4 of MAI (Methylammonium iodide) and DMF (Dimethylformamide), a PCE of more than 10% is achieved. Different solvents, including DMSO and NMP, were also tested for fabricating the active layer. Choosing a suitable solvent is essential for achieving high efficiency. DMSO is one of the mostly used solvents for fabricating perovskite films. PbI_2 has more solubility in DMSO than DMF, because the bond length of Pb–O in DMSO is 2.38 Å as compared to 2.43 Å in DMF. Due to the strong bond between DMSO and lead precursor, deceleration reaction kinetics will happen and will slow down the reaction to the perovskite phase. This will lead to form more uniform crystals with an increase in size. Perovskite layer can be prepared by two methods namely—one-step method and two-step method. Of the two methods, one-step

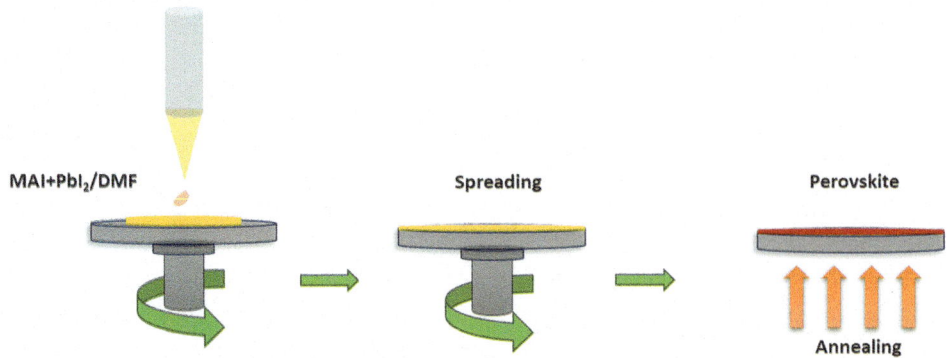

Fig. 5.1 Illustration of one step method

deposition is found to be difficult for coating the perovskite film over the TiO_2 layer. Figure 5.1 depicts the schematic diagram of one step method [1].

Perovskite crystal formation in one-step deposition techniques involves three main stages: (i) the solution reaching the supersaturation stage, (ii) nucleation, and (iii) growth to larger crystals. To achieve nucleation, a supersaturated solution must be present. Upon coating the precursor solution onto the substrate, the solvent quickly evaporates and the solution rapidly gets to saturation (Cs). The nucleation process does not occur at this point because a specific energy barrier must be overcome. With further evaporation, the solution becomes supersaturated with its Gibbs energy greater than the newly formed nuclei, initiating nucleation among atoms, ions, or molecules to form a new phase. Thus, higher supersaturation results in a faster nucleation rate and a higher nucleation density. After the growth of nuclei, crystal growth starts immediately. Eventually, the nucleation slows down and stops because of the decrease in the concentration of the solution due to the continuous formation of nuclei [2, 3].

Studies reported that using an antisolvent over the precursor film increases the crystallization of perovskite. Different antisolvents like CB (Chlorobenzene), toluene, chloroform, and IPA (Isopropyl alcohol) were tested to find the effect of the antisolvent on crystallization. Using antisolvent results in fast nucleation and yields micron-sized nucleates; this method enhanced the efficiency to 14% in planar device structures [1]. Figure 5.2 represents the growth mechanisms occurs during two-step deposition method. It is an alternative method for preparing perovskite materials, which involves depositing lead halide onto the substrate followed by the addition of organic halides in either solution or vapor form. The process of conversion here is driven by a heterogeneous reaction between two layers. During the preparation of $MAPbI_3$ perovskite, two mechanisms may occur: a solid–liquid interfacial reaction under low MAI concentration and a dissolution-crystallization reaction at higher MAI concentrations. An interfacial reaction happens when the concentration of MAI reaches 8 mg mL^{-1} and proceeds according to

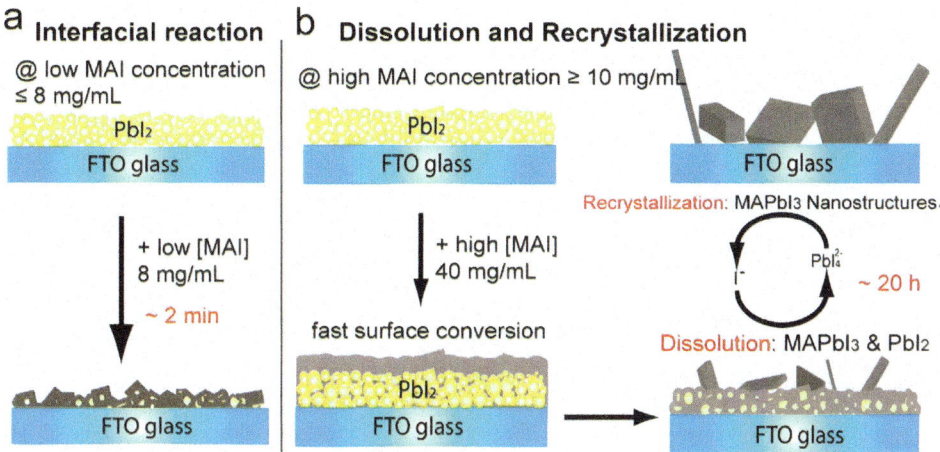

Fig. 5.2 Represents the growth mechanisms occurs in two step deposition process **a** interfacial reaction **b** dissolution and recrystallization (Open access)

the equation

$$PbI_2 + MA^+ + I^- \rightarrow MAPbI_3$$

When MAI concentration exceeds 10 mg mL^{-1}, perovskite crystals were formed due to a solid–liquid interfacial reaction [4]. The reaction between MAI and underlying PbI$_2$ will get suppressed and lead to an incomplete reaction [5]. The incomplete conversion of PbI$_2$ results in structural defects and an elevated charge recombination rate. Fabricating porous PbI$_2$ enhances the contact surface area between MAI and PbI$_2$, addressing the issue of long reaction times and incomplete conversion [2, 6].

5.2 Large Area Fabrication of Perovskite

High efficiency and simple manufacturing of perovskite are the key for commercialization. In the laboratory scale, spin coating is widely used for preparing PSCs. But for large-scale manufacturing, it is difficult for synthesing uniform films using this method. The wastage of precursor solution while using spin-coating technique increases its production cost. Roll to roll method, doctor blade, and spray coating method are commonly used for large-scale manufacturing. But as per previous works, the PCE of PSC is much lesser as compared to the spin coating method. So huge amount of research is going on to find a suitable method for fabricating perovskite layers with uniform morphology and quality [7].

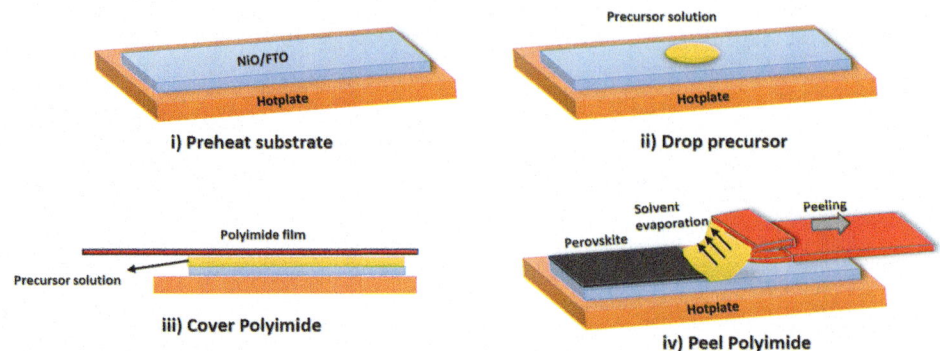

Fig. 5.3 Illustration of the SCD technique

5.2.1 Soft Cover Deposition

Soft cover deposition method (SCD) is considered one of the effective methods for fabricating the active layer. Figure 5.3 illustrates the SCD method. In this technique, a soft film with high wettability, such as PTFE (Polytetrafluoroethylene) or PI is deposited above the perovskite solution drop during the heat treatment process. Preparing perovskite in this method shows rapid evaporation of solvent and fast nucleation, the film deposited is finally peeled off. This deposition method results in high uniformity and high quality. It offers high material utilization of upto 80%, which is more suitable for fabricating large-area perovskite. This method was experimentally done in $MAPbI_3$ and got a PCE of 17.6% for 1 cm^2 cell. For its preparation, the substrate was annealed at 150–270 °C for 30–60 sec and the perovskite solution was dropped above the substrate. After that the substrate was covered with PI film. After 25 s the film was peeled off using a computer-controlled mechanical arm in sidewise at a particular speed.

When the PI film is peeled off fast, solvent evaporation occurs and heterogeneous nucleation happens between the substrate and precursor solution, which results in directional crystallization. This method achieves 100% surface coverage with large grain sizes varying from 500 to 1.2 μm [3].

The SCD method is one among the best methods to fabricate large-scale uniform films economically. Choosing the soft cover, viscosity, and surface wettability of the solution are still some factors that needed more research [2, 8].

5.2.2 Inkjet Printing

The inkjet method is considered an effectual method for fabricating thin films. It works similarly to some inkjet printers, involving computer-controlled film coating. Since it is a non-contact process, no special requirements are needed for the substrate.

5.2 Large Area Fabrication of Perovskite

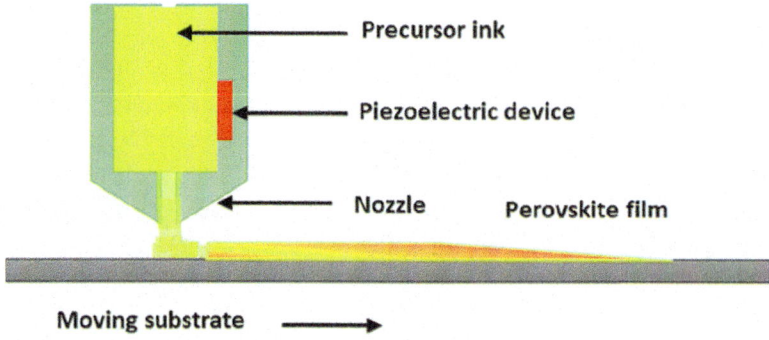

Fig. 5.4 Illustration of inkjet printing

Figure 5.4 depicts the schematic diagram of the inkjet printing process. This technology works based on the piezoelectric effect, the printing process involves three major steps.

(i) The piezoelectric element as shown in the figure shrinks when an electric signal passes through it and creates vacant space.
(ii) The coating solution will rush toward the nozzle.
(iii) The piezoelectric element shrinks again when the droplets are ejected.

A PCE of 11.6% was obtained for a device structure of $TiO_2/CH_3NH_3PbI_3/C$ which was fabricated using the inkjet method in 2014. However, in this study, the perovskite layer is prepared by a two-step preparation technique, where the PbI_2 layer is deposited using the spin-coating method. In recent work, it has been reported that a PCE of 11.3% is obtained in the single-step inkjet method, which can be used for commercialisation purposes. However, the PCE of PSCs prepared using this method is still lower than that achieved by the spin coating technique [2, 3].

5.2.3 Doctor-Blade Coating

The doctor blade method is a simple and economical approach for fabricating films. It is generally used to prepare films in DSSCs, especially for coating the electron transport layer TiO_2. The principle of this method involves spreading the precursor directly onto the substrate using a blade. This method can also be used for large-area films. Figure 5.5 illustrates the different phases involved in the Doctor-blade technique. From the figure, it can be understood that the precursor solution is wiped in a substrate heated to a temperature range of 100–140 °C. The substrate's high temperature speeds up the solvent evaporation and increases nucleation and growth of the perovskite structure.

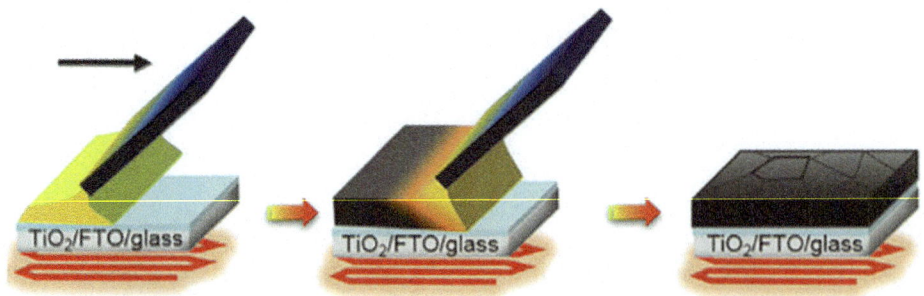

Fig. 5.5 Illustration of various steps in doctor blade method (Copyright)

Compared to the spin-coating method, this method has the facility for applying temperature while coating, thereby controlling the evaporation of solvent. Controlling variables including the solution's surface tension, viscosity, and substrate surface energy, as well as producing a colloidal precursor as opposed to a general one, can further enhance the film's quality [9].

5.3 Dependence of Humidity and Light in the Fabrication of Perovskite

The moisture content present in the environment will affect the growth of perovskite materials like $MAPbI_3$ and $FAPbI_3$. Since perovskite materials are water soluble, the presence of moisture or humidity in the atmosphere inversely affects the film morphology. However, studies found that fabricating perovskite film with 30% humidity improves the growth of the perovskite as well as improves the morphology. But when a fully formed perovskite solar cell is exposed to more than 50% humidity, degradation starts and finally the film will converted into PbI_2. Research found that two types of degradation based on moisture can happen, based on condensation that occurs on the surface of the film and inside of the film. If the perovskite film is exposed to warm humid air more than 50% humidity, degradation occurs due to condensation formed on the surface of film. This causes the film to decompose into HI, PbI_2, CH_3NH_2. This process is irreversible and the device performance is reduced. While exposing the film to cold humid air, no water condensation takes place on the surface, instead, the water gradually integrates into the crystal, forming a distinct crystal structure with isolated $[PbI_6]^{4-}$ octahedra [1]. The resulting crystal structure is also influenced by the temperature, humidity, and duration of exposure.

Firstly, monohydrated crystal phase is formed followed by the dehydrated crystal phase. Water incorporation occurs in an isotropic way, meaning it happens homogeneously all through the sample, rather than starting from the surface and moving inward during crystallization. This process can be reversed and regains its performance by just drying the film in a nitrogen stream.

Moisture during the fabrication affects the performance, morphology and stability of solar cells. A study on the dependence of moisture in the fabrication of $MAPbI_3$ reveals that preparing $MAPbI_3$ with PbI_2 precursor increases stability compared to stoichiometric films. In contrast, the film prepared with excess MAI shows smaller grains and higher disorder in the electronic state. When these samples are exposed to moisture like in the solvent annealing method, recrystallization will occur and form larger highly oriented crystals. The device fabricated from these samples also shows high efficiency.

To some extent, exposure to humidity can improve the device's performance, but prolonged exposure reduces the stability of the material. Studies were conducted on various methods to increase the moisture stability of perovskite films. Making a barrier or encapsulating the perovskite will reduce the moisture-induced degradation. The hole transport material used in PSC also affects its stability. Spiro-OMeTAD is one among the most popular HTMs used in PSCs. However, it forms layers with pinholes on the surface, which poorly protect the perovskite from atmospheric humidity. It also produces PbIOH on the surface of perovskite, accelerating the rate of degradation. Many researches have been conducted to find a suitable HTM that acts as a barrier against moisture and has good performance. Recent studies found that encapsulating a polymer matrix with a layer of Carbon nanotube, can enhance moisture stability and reduce device degradation [1].

The light exposure during the annealing stage of perovskite material has a serious influence on the quality of the films. During fabrication, the material is continuously exposed to light in the laboratory. This study was conducted on the perovskite material $FAPbI_3$, in this study the film was prepared by four different methods to find the effect of light during the annealing stage.

(i) Without dark annealing (The normal method that allow the passage of light during annealing)
(ii) Dark annealing (Blocks the passage of light during annealing)
(iii) Vacuum-assisted without dark annealing (Allows the passage of light but humidity during annealing is controlled by applying vacuum)
(iv) Vacuum-assisted dark annealing (Light and humidity controlled during the annealing stage).

The results show that reducing the exposure of light during annealing results in a much lower bandgap. While the sample that was prepared without light and under a controlled humidity environment showed a significant improvement in the morphology with fewer pinholes and other defects. The perovskite layer prepared doesn't have any sign of delta phase. It shows a wider absorption region and a very low bandgap value of 1.51 eV. Thus, this study proves that, rather than just humidity, controlling light exposure during the annealing stage also influences perovskite film fabrication [10].

5.4 Stability of PSC

Stability, efficiency and lower cost are the major factors that a photovoltaic device must possess. Photovoltaic devices are expected to work for more than 25 years. Si solar cells are one of the devices which have a higher lifetime. In the case of PSCs, various factors affect their lifetime including humidity, temperature, exposure to UV light, and reaction between adjacent functional layers. Thermal stress is also another factor which leads to degradation in perovskite films. Under normal conditions, when the device is exposed to a 40 °C environment, the heat accumulation in the device can increase up to 80 °C. Unlike encapsulation, which solves the moisture stability issue, thermal stability can only be solved by adding cooling support to the device. For MAPbI$_3$, a widely used perovskite material in PSCs, the transition from tetragonal to cubic phase occurs at 57 °C and decomposition begins at 85 °C, even under inert atmosphere. Compared to MAPbI$_3$ the FA-based cells show much more thermal stability. However, a drawback of FAPbI$_3$ is the transition from the cubic phase to the hexagonal delta phase that occurs at room temperature, making it difficult to maintain the black cubic phase. Better stability can be observed for MA or Cs-doped FAPbI$_3$ as a consequence of the strong interaction between FA and Iodine, which decreases its tolerance factor. Even after 500 h of thermal ageing at 85 °C, the FA/Cs-based device retains 90% of its efficiency, indicating its higher stability [8].

Other than the perovskite layer, hole transfer layer used in PSCs also faces stability issues which affects its performance. PTAA and Spiro-OMeTAD are the mostly used HTMs. Both these materials are expensive and that will increase the total cost of the solar cell. Moreover, spiro-OMeTAD needed additional additives like TBP and bis(trifluoromethane)sulfonamide to enhance its electronic properties and stability. TBP enhances the polarity of the HTM and improves the interfacial contact between HTM and the perovskite layer, leading to higher PCE values [11]. Recently, a new Ag-based metal–organic complex was synthesized, which is mostly stable but expensive. Despite its high cost, it can be regarded as a suitable HTM in PSCs for commercialization [8]. Similar to the case of electron transport materials (ETL), TiO$_2$ is the most frequently used ETL in PSCs. However, the main drawback of TiO$_2$ is its stability issue under UV light exposure, as it degrades when exposed to UV, affecting the overall stability and efficiency of the solar cells. Adding an additional layer of GO-Li (graphene oxide) as an interface layer between the TiO$_2$ and perovskite layer improves the stability of the solar cell. Another study, uses ZnO nanorods as ETL instead of TiO$_2$, showing excellent stability and maintaining 90% of its efficiency even without encapsulation [3].

5.4.1 Effect of Encapsulation in PSC

Stability is a major challenge for PSCs. Encapsulating the solar cell helps to protect it from moisture and oxygen. It solves the stability issue of PSC up to some extent. EVA (ethylene vinyl acetate) is one of the materials which is generally used in silicon solar cells. It protects the cell from UV rays, moisture, water, and oxygen. The problem of EVA with PSC is that it is a hot-melt adhesive that requires a temperature of 140 °C for encapsulating. But in the case of PSC, the processing temperature of MAPbI$_3$ is around 100 °C and for FAPbI$_3$ it is nearly 150 °C. This makes the EVA unsuitable for encapsulation in PSC. Encapsulating material should satisfy certain criteria's.

 i. Encapsulants must have high transmittance
 ii. Low cost suitable for commercialization
 iii. Low leakage current to ensure safety
 iv. Capacity to resist any impacts
 v. Ability to protect the PSC from UV radiation, oxygen and moisture.

There are mainly two types of encapsulation approaches edge seal and TFE (thin film encapsulation). Figure 5.6 represents the schematic diagram of (a) Thin film encapsulation method (b) Edge seal method. In TFE, thin films suitable for encapsulation are deposited between the glass substrate and glass cover. The film must be transparent, anti-reflective and impact-resistant. In the edge seal method, the encapsulant is placed around the edges, ensuring that it does not directly contact the perovskite material. This method reduces the chance of any reaction between the encapsulant and PSC [12].

In the case of PSC thermal stress, high pressure and UV curing during encapsulation may affect the PSC performance. Minimising these is the key to encapsulating PSCs. For PSC a temperature of 80 °C is not shown to have a significant effect in the performance and efficiency of PSC. But when the PSC is heated more than 100–120 °C, a significant change in V$_{OC}$ is observed, with no change in fill factor (FF) and current density (J$_{SC}$). A study was conducted to compare the performance of three encapsulants- polyolefin, polyurethane and EVA. Polyurethane was used to encapsulate a PSC of 10 cm × 10 cm, and material silicon was used as an edge seal. To test the performance of encapsulated PSC, the device dipped in water. Even after 2136 h of outdoor testing, PSC retains its 97.52% of initial efficiency. In another study, a composite of PMMA (poly(methylmethacrylate)) and graphene oxide (rGO) is used as an encapsulant, after encapsulation the thermal stability and stability against humidity are increased. The device maintains 90% of its initial efficiency when it is dipped in water for 5 min. Also, there is no noticeable change observed when the device is stored for 1000 h at 35 °C and 40% relative humidity [12].

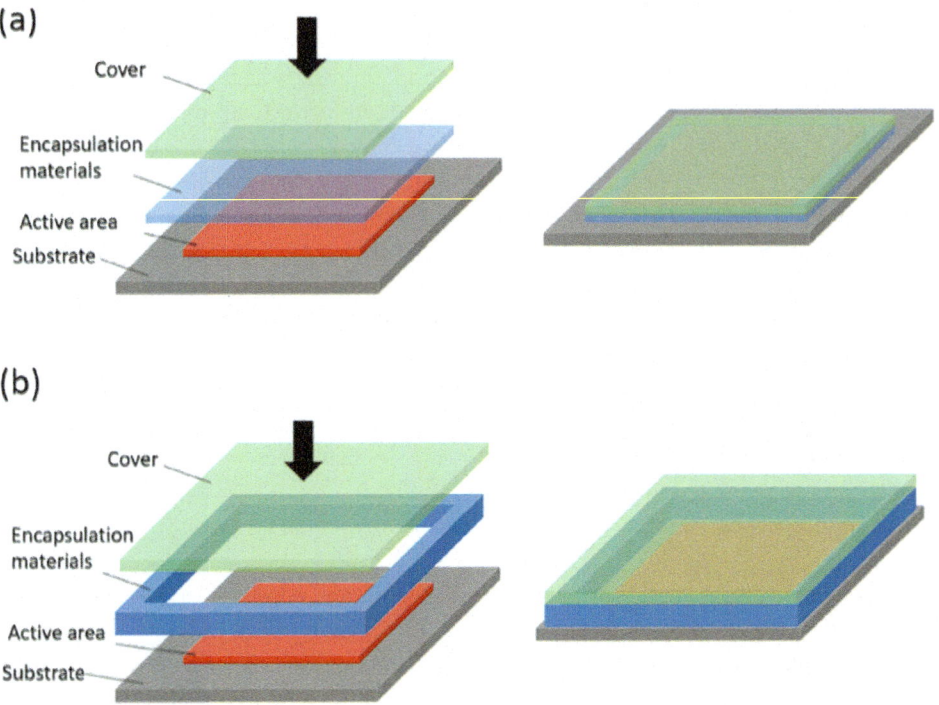

Fig. 5.6 Illustrates the various encapsulation methods **a** thin film encapsulation (TFE) **b** edge seal method (copyright)

Another way of applying encapsulant is through the lamination process, this is widely used in the case of Si solar cells. But in the case of PSC, temperature and the force applied while lamination might damage the perovskite layer. Using a low temperature and low pressure while encapsulating may solve the problem. Other than these issues chance of reacting the encapsulant material with the layers of PSC is also high. It is found that encapsulating PSC with a composite of PMMA and Parylene-C ((poly(p-chloro-xylylene), starts a reaction with spiro-OMeTAD (HTM) and causes degradation. Thus PMMA can't be used to encapsulate PSCs. While, Parylene-C with very low water permeability of 0.14 cm^3 mil/(100 in^2 24 h atm and have high contact angle of 121° can be used. The parylene-C used as an encapsulant in PSC retains an efficiency of 100% even after keeping the device at 26.1 °C and 50% relative humidity [12].

Edge sealing is another way of encapsulation, which protects the PSC from moisture and oxygen. Butyl rubber encapsulation is an effective encapsulation technique that creates a tight seal against air and water. A study uses PIB (Polyisobutylene) a kind of butyl rubber in encapsulating PSC and compared its performance with other encapsulants like UV-epoxy and EVA. Compared to others, the PIB outperforms EVA and UV-epoxy in terms of the protection it offers against moisture and oxygen. It has also a very low glass transition temperature of -75 °C, thus the material can be used at very low-temperature conditions. The encapsulated PSC shows a negligible change in PCE in the damp heat test conducted at 85 °C and 85% humidity for 1000 h. In addition the encapsulated PSC shows a higher efficiency of 120%, even after 200 thermal cycles between -40 and 85 °C. The higher stability and efficiency are due to the blocking of volatilization of decomposed materials from the PSC. Epoxy resin is another material that can be used as edge sealing material, it offers excellent protection against moisture and oxygen. But the curing of epoxy resin is an exothermic process that may affect PSC performance. Comparing with three different methods of encapsulation like AB epoxy glue, UV-curable epoxy and thermally curable epoxy, UV-curable epoxy is the most appropriate for PSC. A recent study reveals that the device exhibits stable performance at 85% humidity when the PSC is encapsulated with UV epoxy along with a desiccant. Another study that used UV-epoxy as an encapsulant, combined with nonpolar paraffin to fill the encapsulation gap, effectively reduced the effect of oxygen and moisture in the PSC. This technique also reduced the escape of volatile components during the decomposition process [12].

5.5 Addressing Toxicity of Lead-Based PSC

All of the high-performance PSCs use Pb-based material as active layers. Compared to cadmium telluride (CdTe) solar cells, perovskite solar cells are much more unstable and easily decompose to PbI_2 when exposed to water. The PbI_2 has a solubility product constant of 9.8×10^{-9}, making it more soluble than the toxic cadmium (Cd) in CdTe solar cells [8]. Since it has a high solubility rate it can easily contaminate soil, ground waters, and rivers. Exposure to PbI_2 primarily leads to severe neurological, cardiovascular, developmental, and reproductive diseases. Thus, these issues of Pb-based solar cells should be dealt with before commercialization [13].

Recently, a study on the environmental effects of PSCs was conducted. In this study, the worst-case scenario of lead exposure was considered, where a damaged Pb-based PSC was completely exposed to a simulated rain device, and the results were analyzed. A PSC module of 300 nm thick contains 0.4 g m^{-2} of Pb. The rain that falls on the Pb device shortly increases the concentration of Pb in the soil to approx. 70 ppm. Eventually, the concentration in soil decreases as Pb spreads across more areas. In uncontaminated soil, the concentration of Pb is approx. 10–30 ppm and 50–200 ppm in higher urban areas. Therefore, the pollution created by these kinds of devices is much smaller (low levels of contamination, < 400 ppm) compared to disasters. However, due to the toxic issue of lead, the release of Pb into the environment must be minimised [8].

Lead-free PSCs are one of the most researched areas today. Considering its spatial arrangement and charge balance, the material that replaces Pb must have a proper ionic size and + 2 charge. In theory, choosing a metal ion with a smaller radius will create strong geometric distortion, resulting in a lower p-p electron transition and a larger carrier effective mass. Tin (Sn) is considered one of the best elements to replace lead (Pb) due to its less toxicity. In 2014, it was reported that PSCs that use $MASnI_3$ show a higher efficiency of 6.4%. Also, it has the added advantage of bandgap tuning. Similar to Pb-based PSC, the bandgap can be tuned from 1.30 to 2.15 eV by substituting I^- with Br^-. However, the fast crystallization rate of Sn-based devices leads to non-uniform films. The partial coverage of the active layer will create a high-charge recombination pathway by making direct contact with ETL and HTL. High surface coverage of the perovskite layer can avoid the ETL/HTL interface and result in higher efficiency. However, controlling the tin-perovskite crystallization is a difficult task due to the rapid reaction between organic and inorganic halides. It is found that using DMSO solvent instead of DMF (dimethylformamide) will result in less pinhole formation and more uniform perovskite films. The solvent DMSO can retard the reaction rate between organic and inorganic halides and can result in higher-quality films with lower pinholes [8, 13].

An alternative way to minimise the toxicity of lead in the environment is recycling. Lead is commonly used in battery and electronic devices, and in the case of PSCs also lead is unavoidable due to its performance. Till now a material which can replace lead is not found. At this stage, recycling is the best method to reduce environmental hazards. Researchers found that we should follow a reclamation program as used in the case of CdTe for lead-based PSCs. The device must be recycled at the end of its life cycle. More than 80% of electronic devices, including batteries contain lead, lead is one of the substances that can be effectively recycled. Other than recycling, robustly sealing, encapsulating solar cells, and applying sodium sulphide to the Pb, which converts soluble Pb to insoluble Pb sulphide are other effective ways to reduce Pb contamination. Rather than recycling Pb, the Pb recycled from batteries can be used to make solar cells. The cost of recycling Pb is significantly lower than the high material costs of electrodes such as Ag, Au, and Cu typically used in solar cells. Based on research about recycling Organic–inorganic halide perovskites (OHP), shows that the solution method can selectively dissolve OHPs and reuse TiO_2 conducting glass. It is found that even after 10 regeneration cycles the initial PCE is maintained. The Au electrode from the PSC can be removed by dipping it in chlorobenzene solution, which dissolves organic hole transport materials. Afterwards, Au having high purity can be separated by washing with acetone, followed by dipping in 20% HCl solution and distilled water. The removal of the perovskite layer can be achieved through a two-step process. First, the MAI layer was removed by dipping it in an ethanol solution, leaving the PbI_2 alone. Then PbI_2 layer was separated by dissolving in DMF. The remaining TiO_2 in the FTO substrate can be removed by ethanol and using UV treatment. The resulting FTO can be directly used without any heat treatment. Researchers reported that the entire process can be done within 10 min and the efficiency remains the same as that is prepared with new materials [8]. In another method of PSC recycling, the perovskite film is heated to a temperature of 250 °C to initiate thermal decomposition. A complete conversation of $MAPbI_3$ to PbI_2 happens at this temperature. The PbI_2 produced is readily used for PSC fabrication and an efficiency similar to pristine cell is obtained. These two methods of recycling PSC are highly valuable since they can reduce hazards caused by the PbI_2 [8].

References

1. M.L. Petrus, J. Schlipf, C. Li, T.P. Gujar, N. Giesbrecht, P. Müller-Buschbaum, M. Thelakkat, T. Bein, S. Hüttner, P. Docampo, Adv. Energy Mater. **7** (2017)
2. F. Huang, M. Li, P. Siffalovic, G. Cao, J. Tian, Energy Environ. Sci. **12**, 518 (2019)
3. Y. Chen, L. Zhang, Y. Zhang, H. Gao, H. Yan, RSC Adv. **8**, 10489 (2018)
4. X.B. Cao, Y.H. Li, F. Fang, X. Cui, Y.W. Yao, J.Q. Wei, RSC Adv. **6**, 70925 (2016)
5. N. Yaghoobi Nia, D. Saranin, A.L. Palma, A. Di Carlo, in *Solar Cells and Light Management*, ed. by F. Enrichi, G.C. Righini (Elsevier, 2020), pp. 163–228
6. P. Zhu, J. Zhu, InfoMat **2**, 341 (2020)
7. M.K. Rao, D.N. Sangeetha, M. Selvakumar, Y.N. Sudhakar, M.G. Mahesha, Sol. Energy **218**, 469 (2021)
8. P. Wang, Y. Wu, B. Cai, Q. Ma, X. Zheng, W.H. Zhang, Adv. Funct. Mater. **29** (2019)
9. J. Li, R. Munir, Y. Fan, T. Niu, Y. Liu, Y. Zhong, Z. Yang, Y. Tian, B. Liu, J. Sun, D.-M. Smilgies, S. Thoroddsen, A. Amassian, K. Zhao, S. Liu, Joule **2**, 1313 (2018)
10. P. Arjun Suresh, G.S. John, A.M. Johnson, U.S. Sajeev, K.V. Arun Kumar, J. Mater. Sci.: Mater. Electron. **35** (2024)
11. T.-T. Bui, F. Goubard, J. Troughton, T. Watson, J. Mater. Sci. Mater. Electron. **28**, 17551 (2017)
12. J. Li, R. Xia, W. Qi, X. Zhou, J. Cheng, Y. Chen, G. Hou, Y. Ding, Y. Li, Y. Zhao, X. Zhang, J. Power Sources **485** (2021)
13. A.B. Djurišić, F.Z. Liu, H.W. Tam, M.K. Wong, A. Ng, C. Surya, W. Chen, Z.B. He, Prog. Quant. Electron. **53**, 1 (2017)

Future of PSC

6.1 Tandem Solar Cells

The current market leader among photovoltaic devices is Si solar cells and it have a laboratory efficiency of 25%. The main reason why it can't reach the ideal efficiency is its thermal losses. In contrast with single junction solar cells, tandem solar cells exhibit higher efficiency due to the utilization of shorter wavelengths of the solar spectrum. If the photons falling on the device have lower energy compared to its bandgap value, they cannot be absorbed while if it is higher, the excess energy beyond the bandgap will dissipate in the form of heat. Figure 6.1 represents four terminal (4T) and two terminal (2T) tandem solar cells. In a Tandem solar cell, the active layer of different bandgaps is stacked one after the other, in such a way that it can absorb maximum radiations from the solar spectrum. The light falling above the bandgap of the top cell is transmitted to the bottom cells. The theoretical efficiency of double, triple, and four junction solar cells is 45, 51, 55% respectively [1].

In the double junction solar cell, the material that has a higher bandgap is stacked on the top and the lower at the bottom, connecting in a way such that each layer works independently. Thus, the total power produced by the solar cell is the sum of two cells. However, in this device, an insulating material is required to separate the two cells. The material is chosen in such a way as to reduce the optical loss from Fresnel reflection [2]. The perovskite tandem cell was first prepared in 2014 with an efficiency of 4.4%. This device is fabricated in the two-terminal configuration, where the top (Perovskite) and bottom (CZTS) layer is connected in series. Later four terminal devices were prepared and published with an efficiency of 25.2% in 2016. The bottom cells used in this device are silicon and it is independently connected to the top cell [1].

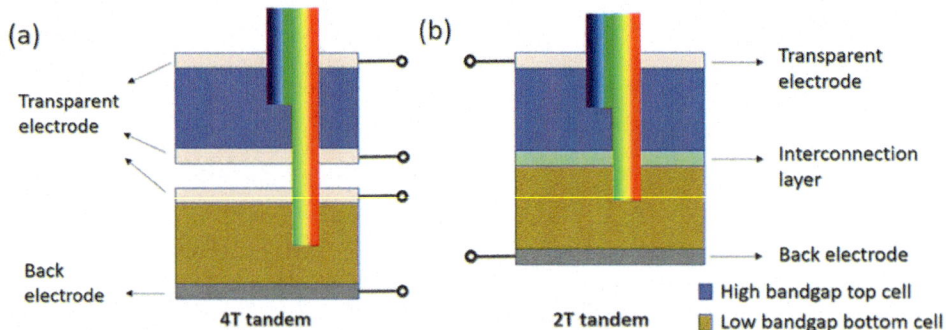

Fig. 6.1 Represents the schematic diagram of **a** four terminal tandem device **b** two terminal tandem device (Open access)

6.1.1 Tandem Cell Configurations

Tandem solar cells can be prepared by four main configurations (shown in Fig. 6.2), by varying the electrical and optical independence.

As shown in Fig. 6.3, in 4T cells, the top and bottom cells are separately connected externally, and the power output is the sum of the individual cells. This is one of the efficient configurations (PCE 25.2%), as current matching between the top and bottom cells is not required. In this type of cell, the top layer is generally perovskite followed by silicon or a CIGS system [1, 3]. For a perfect tandem cell, the top cell must be equipped with highly transparent electrodes on the front and rear. For a 4T configuration, the top cell absorbs radiations from the visible region while the bottom cell absorbs at the near-IR region. Thus the rear electrode of the top cell must have higher transparency in the near infrared region.

ITO (Indium tin oxide), AZO (aluminium-doped zinc oxide), and IZO (indium zinc oxide), are the most commonly used transparent electrodes, and these electrodes are deposited using the sputtering method. 4T cells are usually fabricated on ITO or glass

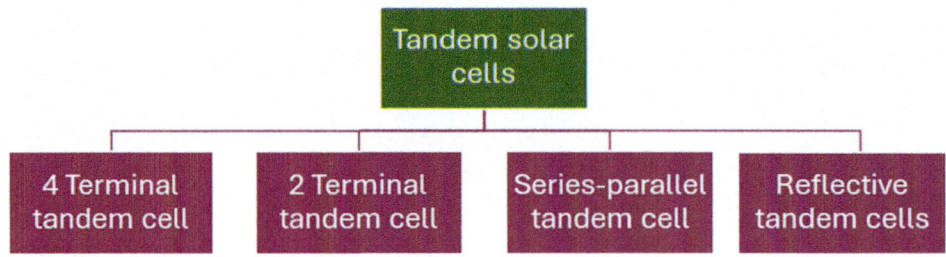

Fig. 6.2 Classifications of tandem solar cells

Fig. 6.3 Tandem configurations of four terminal tandem cell

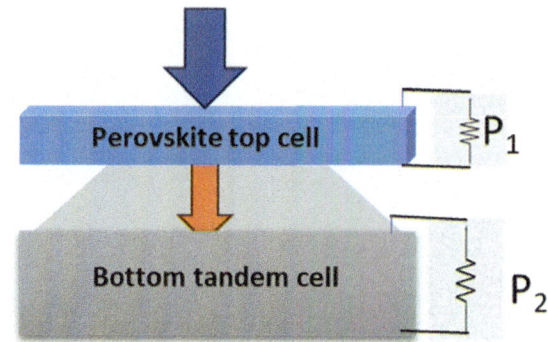

substrates. They consist of a multi-layered ETL, a perovskite active layer typically made of MAPbI$_3$ or FAPbI$_3$ and a top contact with a metallized grid. The bandgap values of 4T are in the range of 1.6–2 eV for the top cell and 1.81 eV for the c-Si bottom cell. The efficiency of perovskite-CIGS tandem solar cells is 22.1%, while perovskite-perovskite-based tandem devices have achieved 20.3%. Reducing the electrical and optical losses in a 4T tandem device can further enhance its efficiency values [4].

In the case of 2T tandem solar cells, the top and bottom layers are stacked one above the other as represented in Fig. 6.4. Compared to the 4T configuration, 2T cells are more complex to manufacture. In a 2T device, the top perovskite cell must match the temperature restriction with the bottom silicon cells. This is done by sensitively controlling the thickness of the absorption layer and carefully designing the integrated structure optically. The advantage of a 2T device is that it requires only one substrate and one external circuit. A transparent conducting electrode with a metal grid and transparent recombination layer is essential for 2T PSCs. The recombination layer used in 2T devices must have certain characterisations,

Fig. 6.4 Tandem configurations of two terminal tandem device

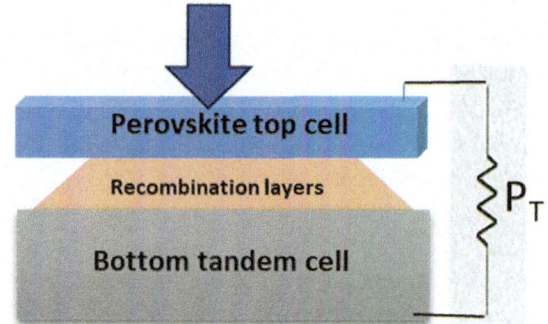

Fig. 6.5 Tandem configurations of reflective tandem cells

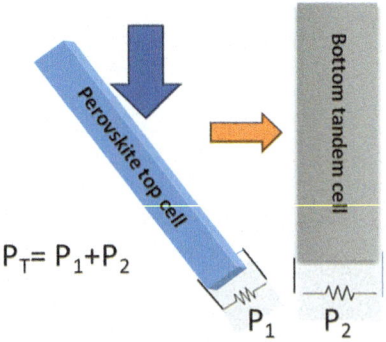

- Excellent transparency
- High carrier recombination rate
- Good ohmic contact
- Inert with other component manufacturing process.

Current matching with the top and bottom layers also helps to enhance efficiency [1, 5]. The highest efficiency of a monolithic perovskite tandem cell with a 2T configuration is 23.6%, which is more than 21.3% of the efficiency obtained for the previous perovskite-silicon configuration using SHJ as the bottom cell [4].

Reflective tandem solar cells have a reflective layer to redirect unabsorbed radiations from the top layer. Figure 6.5 illustrates Reflective tandem cell configurations. In this type of configuration, the top layer is generally perovskite, which has a higher band gap, while the bottom cell has of lower bandgap. The sum of the power generated by both the top and bottom cells gives the total power output of the device. However, the main drawback is the stability of the perovskite layer. Since perovskite is thermally unstable, exposure to higher temperatures may affect its performance [1].

A series–parallel tandem (SPT) PSC (shown in Fig. 6.6) is designed to combine the advantages of series and parallel configurations. In a series configuration, the positive terminal of the first cell is connected to the negative terminal of the other, causes a rise in overall voltage and power output of the device. In contrast, in a parallel connection, all cells are stacked together in such a way that all positive terminals and all negative terminals are separately connected. This configuration makes more current output and also has an added advantage that if any of the cells is damaged, it does not impact the device performance. Blending the benefits of series and parallel connections, an SPT cell can balance voltage and current outputs. This configuration can achieve high PCE and improved performance [4].

Fig. 6.6 Tandem configurations of series–parallel tandem cells

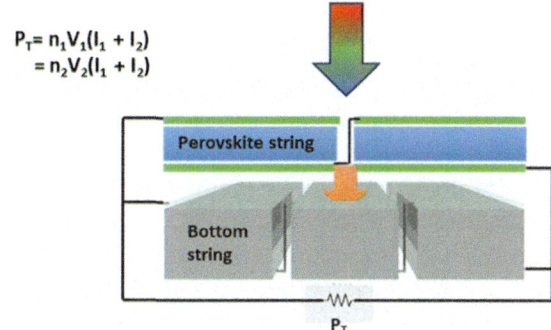

6.2 Pros and Cons of Perovskite Material in Tandem Solar Cells

Perovskite materials exhibit various advantages as well as challenges in the area of tandem solar cells. Flexibility, tunable bandgap, and high photovoltaics are its main benefits while the challenges include, their lack of stability and complexity in film deposition techniques.

Tunable bandgap
Inorganic–organic perovskite materials can tune the bandgap within the range of 1.2–2.2 eV by changing the composition of A, B and X ions. Varying the A site ions will distort the perovskite structure and it leads to a change in the bond length and angle between B and X. Sn^{2+}, Ge^{2+}, and Pb^{2+} are the most commonly used B cations, which have a bridging angle of 155.2°, 166.3° and 159.6° respectively with the BI_6 octahedra. When the bridging angle increases, the bandgap of the material decreases: $APbX_3$ > $ASnX_3$ > $AGeX_3$. In a similar way, when the electronegativity of the X halogen anion increases, the ionic character of the B-X bond increases due to strong electron attraction by the halogen. This results in a higher bandgap. Thus we can control the bandgap of a perovskite material by varying A, B and X ions. For the majority of devices having higher efficiency, the desired bandgap and stability are achieved by changing A, B and X ions. The easy way of modifying the bandgap is by utilizing the halide's tunability between bromine and iodine. In both methyl ammonium and formamidinium lead halide systems, the bandgap can be varied from 1.55 to 2.0 eV by changing the ratio of Br and I. It is very difficult to lower the bandgap to less than 1.5 eV, especially while substituting Pb with Sn. Adding tin to the perovskite structure causes serious stability issues. Tin exists in the Sn^{2+} state, which is required for semiconducting properties. However, due to stability issues the Sn^{2+} will get oxidised to Sn^{4+} when it is exposed to moisture or oxygen. These

Sn^{4+} will act as p-type dopants and limit the device's performance and stability. A single-junction solar cell uses a rear reflector with a thickness at the micron level to achieve higher efficiency, but in the case of tandem cells, the top cell must transmit sub-band light into the bottom solar cell [1, 4].

Photovoltaics of perovskite materials
In the case of tandem solar cells, higher efficiency would be obtained only if we can extract an electrical current of high voltage from the top cell than from the bottom cells. This demand the top cell to have high radiative efficiency, a property for which organic–inorganic halide perovskites are well known, exhibiting a photoluminescence radiative efficiency of $\varphi = 0.70$. The higher photoluminescence efficiency of perovskite materials will lead to a low voltage difference between bandgap energy and V_{OC} in comparison with other thin film solar cell materials. The smaller voltage difference indicates that the absorbed light is converted into electrical energy more efficiently, which makes perovskite material more beneficial for photovoltaic applications.

Stability of perovskite materials
Since perovskite is used as an active layer in tandem solar cells, the stability of PSC influences the device's performance. The stability of perovskite relies on various factors including the preparation technique, composition of perovskite, HTL and ETL materials, metal contacts, and encapsulation methods. Since it is a tandem device, the transparency of electrodes has a major impact on efficiency. It has been found that using sputtered ITO instead of thermally evaporated electrodes increases the thermal stability of perovskite solar cells. ITO electrodes provides better protection for perovskite layers against moisture and reduces the corrosion of metal electrodes.

Transparency of layers
Tandem solar cells are devices that have two solar cells combined together. Thus, the transparency of the top cell matters the overall output of tandem solar cells. This includes the transparency of contacts, charge selective layers and recombination layers. But the transparency of the top cell also influences its bandgap of a perovskite cell. Thus, a high-efficiency tandem solar cell, can selectively control the wavelength in such a way that the top layer absorbs shorter wavelength and the bottom layer absorbs higher wavelength [1].

Recent single-junction perovskite solar cells use a reflector in the rear of the device, which improves efficiency by increasing the path length of light through the active layer at least two times. This configuration maximizes light absorption and boosts the generation rate of electron–hole pairs in the active layer. Since the transparency of the device is crucial for higher efficiency in tandem solar cells, the rear reflector arrangement must be replaced with a transparent one. This adjustment decreases the absorption in the perovskite layer compared to single-junction PSCs. Thus, further investigation is needed to solve these issues and increase its efficiency [4].

Transparent electrodes and recombination layers

In the case of tandem devices, they must exhibit higher transparency and minimal electrical losses. For a 4T solar cell, both the back contact of the top cell and the front contact of the bottom cell must be transparent to sub-bandgap light in the N-IR region. Also factors such as, band-gap alignment, spatial localization in the tandem stack, sensitivity to temperature and impacts must be considered while selecting contacts. The transparent conductors must have sheet resistance below 10 Ω sq^{-1} and have a transmittance of more than 80% for 400–1100 nm wavelength region for maximum device performance. Furthermore, the rear electrode must be minimised to get higher efficiency. The rear reflection from the solar cell may result in interference of light, leading to total reflection or transmittance [1, 6].

In 2T tandem cells, designing and fabricating the recombination layer is of great significance. The recombination layer must have a high transparency rate while minimizing voltage loss to ensure optimal device performance. In silicon-perovskite tandem devices, transparent conducting oxides are commonly used as recombination layers. As part of earlier research in 2016, Eperon et al. used ITO-based material as recombination layers. In this device, SnO_2/PCBM is used as ETL while PEDOT: PSS is used as HTL. A 4 nm thick tin oxide and 2 nm layer stack were deposited above the PCBM layer by the ALD method (Atomic layer deposition), which acts as a buffer layer during ITO sputtering. The ITO layer not only facilitates charge recombination but also shields the perovskite layer from humidity. Today, Indium-doped materials are mostly used for recombination layers due to their compatibility with tandem devices [4, 7].

6.3 Efficiency Progress of Tandem Solar Cells

Theoretically, tandem solar cell is able to attain an efficiency of up to 47%, which is greater than the SQ limit for single junction devices under non-concentrated light (AM 1.5G spectrum). Perovskite materials having bandgap of 1.55–2.2 eV can be employed as the top layer of tandem solar cells along with lower bandgap cells like Crystalline silicon or CIGS. Recent perovskite materials like $FA_{0.75}Cs_{0.25}Sn_{0.5}Pb_{0.5}I_3$ and $MA_{0.5}FA_{0.5}Pb_{0.75}Sn_{0.25}I_3$ having a low bandgap value of 1.2 eV and 1.33 eV can also be employed as bottom layer for tandem solar cells. Thus, PSCs can be even used as a lower bandgap material for 4T and 2T tandem solar cells. CZTS-perovskite solar cell was the first tandem solar cell reported with a PCE of 4.4%, the reason for low efficiency is due to the opaque nature of thin aluminium contact. The efficiency was further increased to 25.2% for four terminal devices which use silicon and CIGS as bottom devices [1].

As shown in Fig. 6.7 the top layer of four-terminal tandem solar cells is typically fabricated on an ITO substrate followed by a multi-layered ETL, perovskite layer like $MAPbI_3$ or $FAPbI_3$, HTL such as Spiro-OMeTAD, MoO_3 buffer layer and metal gridded ITO top contact.

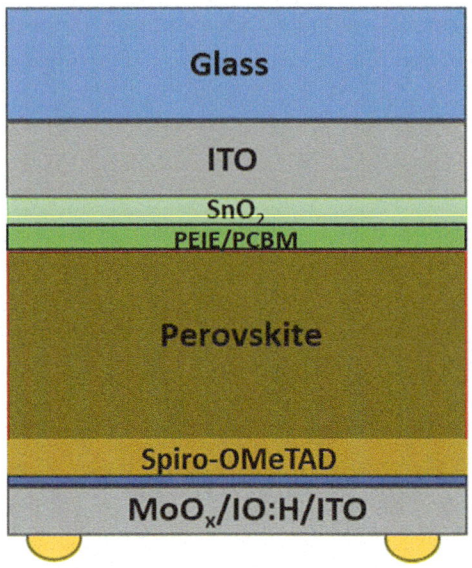

Fig. 6.7 Illustrates a schematic diagram four-terminal tandem device

In 2T tandem solar cells (as illustrated in Fig. 6.8), the current efficiency limit is nearly 23.6%, which is certified by NREL. These cell uses SHJ (silicon heterojunction) as the bottom cell and IZO as the recombination layer. It facilitates efficient charge recombination between the p-type hole transporting layer of Si solar cell and ETL of the perovskite cell [8]. ETL used here is a blend of [6,6]-phenyl-C_{61}-butyric acid methyl ester (PCBM) and polyethyleneimine (PEIE) [1, 4].

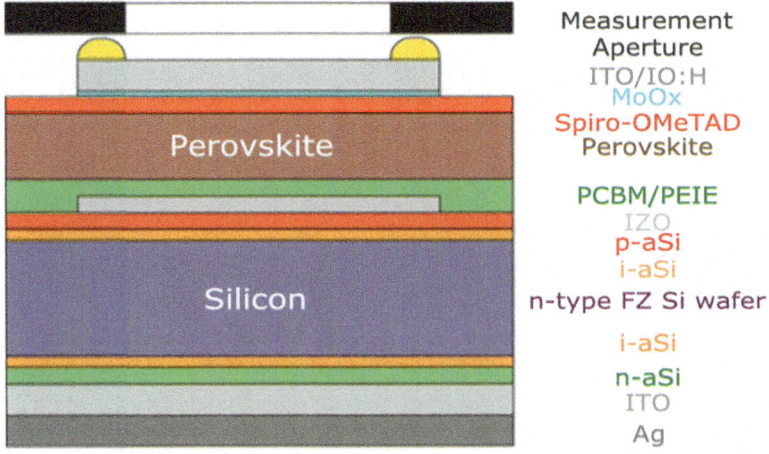

Fig. 6.8 Schematic diagram of 2T perovskite/Si solar cells (Open access)

6.4 Flexible Solar Cells

PSCs are futuristic solar cells which offers excellent efficiency, simple preparation and low cost. It is generally prepared in a rigid glass substrate like in silicon solar cells. Studies have found that due to inherently flexible mechanical properties along with low-temperature preparation of perovskite, make them suitable for manufacturing flexible solar cells. Flexible properties of substrate, low-temperature processed HTL and ETL are some of the factors that want to be considered while preparing flexible solar cells [9, 10].

6.4.1 Flexible Substrates

The selection of a suitable substrate is one of the important aspects to consider while preparing flexible PSCs. The deposition of adjacent layers is very much relied on the mechanical and chemical characteristics of the substrate. An ideal substrate should have high transparency, high-temperature tolerance, high mechanical strength, resistance to chemical degradation, behave like a barrier against oxygen and moisture, and low surface roughness. However, till now, no single substrate has been able to satisfy all these requirements. Glass substrates, metal substrates and polymer substrates are the most commonly used flexible substrates [10, 11].

6.4.1.1 Flexible Glass Substrates

Glass substrates that have thicknesses in the micrometre range are considered flexible glass substrates. The first used glass substrate for flexible PSC is willow glass, which has a thickness of 50 μm. This type of FPSC achieved an efficiency of 12.06%. It shows some degradation of 0.5% after 200 bending cycles. Smooth indeformable surfaces and temperature-withstanding capability of the glass substrate facilitate large-area fabrication of FPSC. Recently, FPSC with a size of 42.9 cm^2 was fabricated using the blade coating method and achieved a record efficiency of 15.86%. However, the high cost and fragile nature of glass plates make it difficult for practical applications [10].

6.4.1.2 Metal Foils as Substrates

The high thermal resistance, flexibility, and conductivity of metal foils make them suitable candidates for preparing FPSC. However, due to their opaque nature, they can only be used as bottom electrodes, which is a major drawback in fabrication. Initially, thin Ti foils are used as flexible substrates, allowing the entire device to be fabricated without ITO or FTO layers. The FPSC device structure consists of Ti/TiO$_2$/perovskite material/Spiro-OMeTAD/Ag. The high-temperature resistance of Ti foils allows the TiO$_2$ layer to properly crystallize. In another study, ultrathin Ag is used as a substrate with a thickness of 12 nm to increase optical transmittance. However, the study found that a very thin layer of Ag offers minimal conductivity. Increasing the thickness of Ag will lead to high conductivity, but significantly affects light transparency. The transmittance of the optimal Ag electrode is 45% lower compared to the FTO electrode, achieving the best efficiency of only 6.15%. Even though metal foils have high conductivity, the substrate which has an optimum balance between conductance and transparency is yet to be found [10].

6.4.1.3 Polymer Substrates

Each material has its advantages and disadvantages and this is also true for polymer substrates. Figure 6.9 represents the schematic diagram of flexible polymer solar cells. Polymer materials like PET (Polyethylene Terephthalate), PEN (Polyethylene naphthalate), and PI (Polyimide) are the most commonly used flexible substrates. These polymers are favoured for their low cost, lightweight, high bendability, corrosion resistance, and optical transparency [12]. However, because of their lower thermal stability, fabricating FPSC is a challenging task. The transition temperatures of commonly used substrates—PET, PEN, and PI substrates are 78 °C, 120 °C, and 200 °C respectively [13]. Thus when the temperature exceeds the transition temperature, deformation will occur, affecting the device's structural integrity. Additionally, the resistance of polymer substrates increases significantly when indium tin oxide (ITO) is deposited at higher temperature of 200 °C. As a result, the entire fabrication process of FPSC should be conducted at low temperatures. Since the annealing temperature of $MAPbI_3$ perovskite is 150 °C, efforts are going on to prepare high quality charge transport layer at a lower temperature [10]. Another major drawback of polymer substrates is their poor barrier properties against water and oxygen compared to glass and metal substrates. Despite this drawback, polymer substrates remain prominent due to their excellent performance and ease of fabrication [10, 14].

6.5 Inverted Perovskite Solar Cells

Inverted perovskite solar cells (IPSCs) offer high operational stability, low hysteresis, and allow manufacturing at low temperatures. Figure 6.10 illustrates the layers of inverted PSCs. PSCs are classified into n-i-p (regular) and p-i-n configuration (inverted). The regular configuration of PSC can also be classified into planar and mesoporous PSCs. In a regular PSC with an n-i-p configuration, the perovskite material absorbs the incident light through the electron transport layer (ETL). Early studies indicated that the regular configuration of PSC typically employs mesoporous TiO_2 as the ETL, which necessitates high temperatures for fabrication. When the thickness of the mesoporous layer is decreased, the device encounters significant hysteresis. These fabrication challenges increase the production cost. However, IPSC works oppositely compared to regular configuration [15].

The main steps of working of IPSC are.

i. Incident light enters the perovskite layer through HTM and absorbs photons to form electron–hole pairs, later separated into free carriers.

6.5 Inverted Perovskite Solar Cells

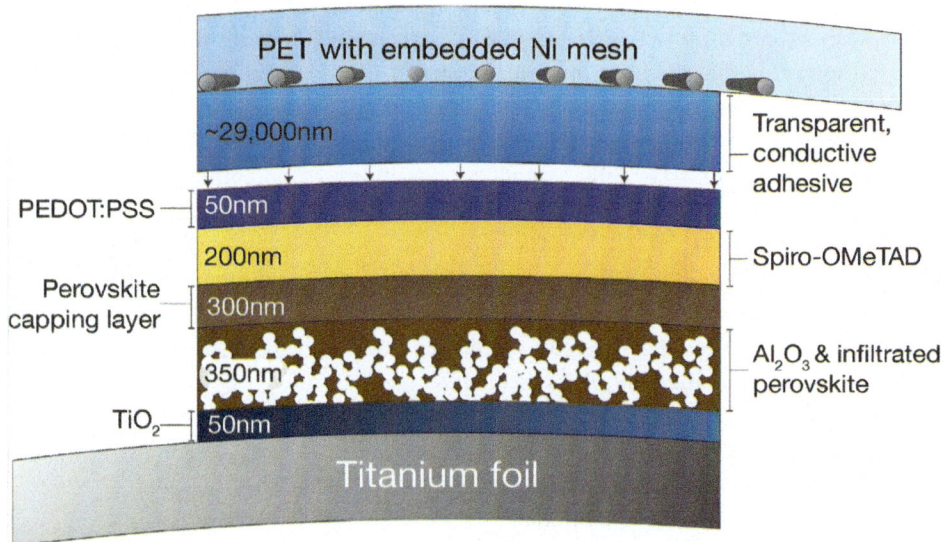

Fig. 6.9 Flexible solar cell with polymer substrate (Copyright)

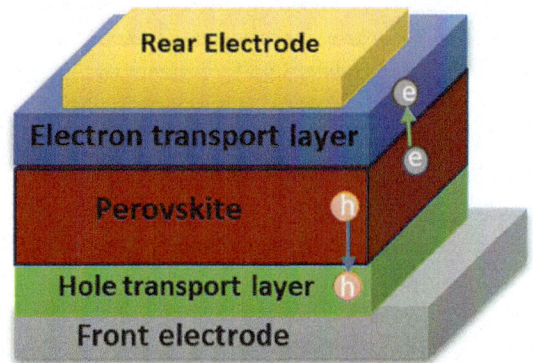

Fig. 6.10 Illustrates the schematic diagram of IPSCs

ii. The electrons created are shifted to the rear electrode through ETM while holes are transported to the front electrodes through HTM.
iii. The charge carriers collected in the electrodes create current in the loop.

The first IPSC was proposed in 2013 with the configuration ITO/PEDOT: PSS/MAPbI$_3$/PCBM/BCP/Al, and this IPSC achieved a PSC of 3.9%. The efficiency obtained was very poor compared to regular structured PSC. Later, a PCE of 15% is gained for IPSC of size 1 cm^2 and exhibited great improvement in stability up to 1000 h in lighted conditions. Apart from that, it shows.

i. Enhanced stability against light and humidity due to the PCBM (1-(3-methoxycarbonyl)-propyl-1-phenyl-(6,6)-C_{61}) layer, which is more hydrophobic compared to HTMs that are used in regular PSCs.
ii. The enhanced electron conductivity of the material, compared to the generally used ETL TiO_2, reduces the hysteresis.
iii. Inverted perovskite solar cells extend the possibility of combining them with other PSC or silicon solar cells to create tandem solar cells having PCE greater than 40%.

Recently, IPSC achieved an efficiency of 22.3%, and the mini-module of IPSC passed the damp heat and humidity test of temperatures up to 85 °C and 85% RH. These results suggest that IPSCs have the potential to replace conventional Si solar cells and could facilitate the commercialization of PSCs [15].

6.5.1 Different Configurations of IPSC

6.5.1.1 Single-Junction Inverted Perovskite Solar Cells

Single-junction solar cells are devices that operate based on a single active layer. In the case of IPSCs, the first device uses a planar heterojunction, which features a donor–acceptor interface and simplifies the fabrication process by eliminating the need for heat treatment to create mesoporous TiO_2. Due to better open-circuit voltage and energy level matching, using NiO_x instead of TiO_2 is more suitable. The device that uses ITO/compact NiO_X/NiO_X nanoparticles/$MAPbI_3$/PCBM/BCP/Al achieved an efficiency of 9.15%. However, the large intrinsic light absorption coefficient and low hole mobility have hindered the further development of IPSCs based on NiO_x. Introducing Al_2O_3 into the NiO_x-perovskite interface reduced the pinholes in NiO_X without affecting the light absorption coefficient. Additionally, the fill factor (FF) of the device increased from 62 to 73%, and the PCE exceeded 13% upon the introduction of Al_2O_3 [15].

6.5.1.2 IPSC Tandem Solar Cells

The PCE of single-junction IPSCs is limited by two major factors: reduced photon absorption and thermal relaxation of hot carriers, as described by the SQ limit. Tandem cells are two junctional devices that use two active layers of suitable bandgap to obtain more efficiency. Among four-terminal and two-terminal tandem devices, the two-terminal device is much more economical due to less number of electrodes and fewer window layers. Sharp absorption edge, variable bandgap, and favourable V_{OC} of 1.15 V make perovskite material a suitable material to combine with proven silicon solar cells. Low-temperature-processed IPSCs with stable and transparent ETLs, along with high thermal stability, are most suitable for use as top cells [15].

A two-terminal IPSC with the configuration perovskite/(Cu(In, Ga)Se_2) achieved an efficiency of 22.43% in 2018 by optimizing the transparency and electrical connection of the interconnecting layer. The surface roughness of the boron-doped ZnO/Cu(In, Ga)Se$_2$ was reduced from 250 to 40 nm through sputtering and chemical polishing. Therefore, for IPSCs, an organic charge transport layer like PEDOT: PSS is preferable [15].

6.5.1.3 Perovskite Layer Fabrication in IPSC

Since perovskite materials serve as light-harvesting layers in PSCs, their uniformity and quality directly impact device efficiency. Producing a uniform perovskite film prevents unwanted contact between the top and bottom charge transport layers Moreover, film nonuniformity reduces the V_{OC} and fill factor (FF). The initially used MAPbI$_3$ has now been replaced with FA-based materials, comprising double-cation and triple-cation mixed halide perovskites such as FAMA and (Cs) FAMA. The fabrication method is also changed to suit for fabricating different compositions of perovskite materials. The one-step method is commonly employed for IPSC fabrication, as it readily produces dense and uniform films. While two-step method is not commonly used, since the formation of perovskite completely depends on the fabrication conditions. Also, the film formed by this method is usually very rough, which is not suitable for coating thin PCBM layers that fully cover the active layer. Other methods like blade coating, vacuum deposition, and slot-die coating are also used for IPSC manufacturing.

MAPbI$_3$ was the most commonly used material for fabricating IPSCs. However, it faces significant thermal stability issues, making it unsuitable for practical applications. Considering thermal stability, FA-based materials perform better due to their strong hydrogen bonding with PbI$_6$ octahedra. Additionally, the reversible decomposition of MA-based devices begins at 85 °C. The narrower bandgap and broader absorption region of FA$^+$ cations, compared to MA$^+$ cations, increase the short-circuit current (J_{SC}) of IPSC. In one study, partially replacing MA with FA in the composition FA$_{0.85}$MA$_{0.15}$Pb(I$_{0.85}$Br$_{0.15}$)$_3$ improved the J_{SC} by 1 mA cm^{-2}, compared to MAPbI$_3$ devices. Furthermore, partially replacing MA$^+$ with Cs$^+$ enhanced film quality by reducing the crystallization rate. Pure FAPbI$_3$ is commonly used in high-performance PSCs due to its narrow bandgap and higher absorption region. However, in the case of inverted PSCs, it is not commonly used due to its difficulty in crystallizing α-FAPbI$_3$ and due to its poor phase stability. Therefore, various research is going on to increase the crystallinity and to stabilise the black alpha phase. MACl is one of the commonly used additives to boost the crystallinity of the perovskite and to achieve a favourable black phase by forming an intermediate phase. This technique is already used in regular PSC and has achieved good results [16].

6.5.1.4 HTMs and ETMs in IPSC

HTM and ETM significantly influence the performance of solar cells. They separate and transport the electrons and holes produced by the perovskite layer to the corresponding electrodes. In IPSC the HTMs are positioned between the perovskite layer and TCO (transparent conducting oxide), while ETM is generally coated above the perovskite layer. The main function of ETMs is to extract electrons and block holes from the perovskite

Fig. 6.11 Energy levels of different components IPSC (Copyright)

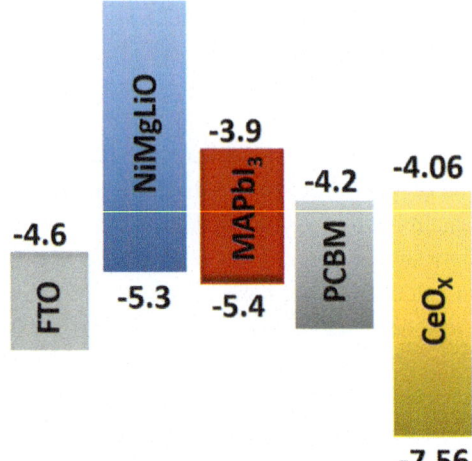

layer while HTMs block electrons and transport holes. Figure 6.11 represents the energy level diagram of commonly used components in IPSC. Charge transport materials (CTMs) effectively separate charge carriers (holes and electrons) by properly aligning their energy levels with those of the perovskite layer.

In the case of IPSC, the hole transport layer directly influences the efficiency and stability of the PSC. The features that an ideal HTM should have are.

i. HTM should have high optical transmittance.
ii. The HOMO level of HTM should align with the VBM of the perovskite layer, to reduce interfacial recombination.
iii. Have high hole mobility, which minimises the transport resistance and will increase the overall efficiency.

HTMs can be classified into organic and inorganic HTMs. For IPSCs, organic HTMs are most widely used, due to their tunability and low-temperature processing. Polymer HTMs come under organic materials [17]. Mostly used polymer HTMs are PEDOT:PSS, PTAA ((2,4,6-tri-methylphenyl)) amine, and poly-TPD (poly(bis(4-phenyl)). However, PEDOT: PSS have some problems since it is highly acidic and has energy level mismatch between the perovskite layer. The perovskite layer will eventually react with PEDOT: PSS's acidic nature, and deteriorates the device performance. In otherwords, poly-TPD suffers from poor wettability and is expensive.

PTAA is also expensive and has poor wettability. However, its suitable energy alignment with the perovskite layer, higher conductivity, optical transmittance, and neutral nature make it a viable choice for enhancing IPSC performance. Devices using PTAA have achieved significant improvements, with a PCE of 25%. These factors hinder their commercialization.

Organic HTMs face challenges such as hydrophobicity and degradation issues. In contrast, inorganic HTMs, which can be used as hole transport layers in IPSCs, offer higher thermal stability and longer lifetimes. Considering different organic materials, NiO_X is the material that is generally used in IPSCs due to its favourable energy level alignment. In undoped NiO_x, the high ionization energy of Ni vacancies reduces the hole charge density, leading to decreased conductivity [18]. Doping with metal ions such as Ag^+, Cu^{2+}, and Au^+ can enhance conductivity by altering the Ni concentration. Recently, doped composite, $Li_{0.05}Mg_{0.15}Ni_{0.8}O$ film showed a conductivity of 2.32×10^{-3} S cm^{-1}, which is more than 12 times that of undoped [19]. Although NiO-based systems demonstrate relatively high efficiency, it is not near to the efficiency obtained by organic HTMs.

Electron transport materials placed above the perovskite layer take a significant part in determining the device's efficiency. Fullerenes and their derivatives are extensively explored in high-performance inverted PSCs because of their high charge mobility and low-temperature synthesising characteristics. Thus, more studies are going on in inorganic materials to increase the performance and stability of IPSC. Due to its improved stability and higher electron mobility (10^3 cm^2 kV^{-1} s^{-1}), CeO_x (cerium oxide) has gained attention in IPSCs. Additionally, its appropriate energy level alignment promotes efficient charge extraction, making it a suitable choice for these devices. The IPSC-based CeO_x gained an efficiency of 18.7% with the configuration, $NiMgLiO/MAPbI_3/PCBM/CeO_X$. It also shows increased stability, retaining 91% of its initial efficiency when exposed to an atmosphere with 30% relative humidity and continuous light exposure. Even though fullerenes meet certain criteria for HTM, it is still expensive, and more research is needed to find a material that has higher efficiency and low cost [18].

6.6 Lead Free PSC

Even though PSCs with relatively high stability show high-performance many researchers have taken into account the presence of lead as a threat, which is to overcome by replacing it with alternative materials such as Bismuth (Bi), Tin (Sn), Germanium (Ge), and Copper (Cu) [20]. Sn is one of the material which can replace lead in the PSCs. Incorporation of Sn in PSCs is typically achieved through the alloying of Sn/Pb, with this device reaching an efficiency of 15% in an oxygen-deficient environment. In a recent study, the Pb:Sn ratio was varied to investigate the performance of perovskite solar cells. It could be observed that Sn is a suitable material that can be used in active layers, but in pristine form, it does not exhibit any photovoltaic properties. It has been proved that Sn inherently maintains its 2+ oxidation state when Pb is present in the absorber layer. In another study, a completely lead-free PSC was fabricated using $MASnI_3$ as the active layer, with encapsulation, under a nitrogen glove box to prevent degradation. However, stability issues were reported for the Sn-based active layer, as discolouration occurred shortly when the absorber layer was exposed to ambient air. This was attributed to the intrinsic property of

Sn^{2+} to rapidly oxidise to Sn^{4+} with the presence of oxygen. To address this degradation issue, the complete cell was fabricated under a nitrogen atmosphere [20].

The researchers compared the characteristics of Sn and Pb-based methylammonium (MA) and formamidinium (FA) halide perovskites. They found that the prepared perovskites exhibited bandgaps varying from 1.1 to 1.7 eV and also demonstrated outstanding photoluminescence characteristics. By the simultaneous application of Pb and Sn ($MAPb1_{-x}Sn_xI_3$) in the absorber layer, the bandgap does not consistently change with the increase in substitution rate. The bandgap of the perovskite material decreases at lower concentrations of Sn and reaches its minimum value at Sn concentrations of $x = 0.55$ and 0.75. When the concentration of Sn is further increased the bandgap value reaches a maximum of 1.30 eV. This trend can be seen in FA and MA halide perovskites [20].

Cs halide perovskites have been studied with Sn to assess their compatibility. The $CsSnI_3$ perovskite absorber can utilize photons in the N-IR region and has a short-circuit photoelectric current density of 34.3 mA cm^{-2} with a lower energy bandgap of 1.3 eV. It was reported that doping $CsSnI_3$ with 20% SnF_2 enhanced its metallic nature by reducing the background current density. When Ge-based perovskite has been studied based on density functional theory, it was found to be suitable to be an absorber layer for photovoltaic solar cells. The performance of the Ge-based perovskite on the absorber layer was studied after it was prepared with various cations such as Cs, FA, MA, guanidinium, isopropylammonium, acetamidinium, and trimethylammonium. However, the Ge-based perovskite achieved only 0.2% efficiency and performed poorly in V_{oc}. In contrast, Cu-based perovskites demonstrated greater stability in air. These 2D perovskites feature multi-quantum well electronic structures and possess unique dielectric, magnetic, and electronic properties [20].

References

1. N.N. Lal, Y. Dkhissi, W. Li, Q. Hou, Y.B. Cheng, U. Bach, Adv. Energy Mater. **7** (2017)
2. A.W.Y. Ho-Baillie, J. Zheng, M.A. Mahmud, F.J. Ma, D.R. McKenzie, M.A. Green, Appl. Phys. Rev. **8** (2021)
3. F. Khan, B.D. Rezgui, M.T. Khan, F. Al-Sulaiman, Renew. Sustain. Energy Rev. **165**, 112553 (2022)
4. S. Ašmontas, M. Mujahid, Nanomaterials **13** (2023)
5. Q. Wali, N.K. Elumalai, Y. Iqbal, A. Uddin, R. Jose, Renew. Sustain. Energy Rev. **84**, 89 (2018)
6. Q. Wang, A. Abate, in *Emerging Photovoltaic Materials* (2018), pp. 261–356
7. S. Seo, S. Jeong, H. Park, H. Shin, N.-G. Park, Chem. Commun. **55**, 2403 (2019)
8. H. Wu, Z. Sun, H. Li, X. Chen, W. Ma, S. Li, Z. Chen, F. Xi, Energy Mater. Devices **2** (2024)
9. F. Di Giacomo, A. Fakharuddin, R. Jose, T.M. Brown, Energy Environ. Sci. **9**, 3007 (2016)
10. G. Tang, F. Yan, Nano Today **39** (2021)
11. M.Z. Qamar, Z. Khalid, R. Shahid, W.C. Tsoi, Y.K. Mishra, A.K.K. Kyaw, M.A. Saeed, Nano Energy **129**, 109994 (2024)

References

12. X. Li, T. Wang, L. Yang, B. Dong, Y. Li, L. Li, L. Li, S. Feng, G. Chen, Y. Yang, J. Energy Chem. **104**, 254 (2025)
13. Q. Wei, W. Zi, Z. Yang, D. Yang, Sol. Energy **174**, 933 (2018)
14. Z. Fakharan, A. Dabirian, Curr. Appl. Phys. **31**, 105 (2021)
15. X. Lin, D. Cui, X. Luo, C. Zhang, Q. Han, Y. Wang, L. Han, Energy Environ. Sci. **13**, 3823 (2020)
16. S. Liu, V.P. Biju, Y. Qi, W. Chen, Z. Liu, NPG Asia Mater. **15** (2023)
17. H.D. Pham, L. Xianqiang, W. Li, S. Manzhos, A.K.K. Kyaw, P. Sonar, Energy Environ. Sci. **12**, 1177 (2019)
18. E.A. Nyiekaa, T.A. Aika, P.E. Orukpe, C.E. Akhabue, E. Danladi, Heliyon **10** (2024)
19. Z. Yu, L. Sun, Small Methods **2**, 1700280 (2018)
20. M. Abd Mutalib, N. Ahmad Ludin, N.A.A. Nik Ruzalman, V. Barrioz, S. Sepeai, M.A. Mat Teridi, M.S. Su'ait, M.A. Ibrahim, K. Sopian, Mater. Renew. Sustain. Energy **7** (2018)

The manufacturer's authorised representative in the EU is Springer Nature Customer Service Centre GmbH, Europaplatz 3, 69115 Heidelberg, Germany. If you have any concerns regarding our products, please contact ProductSafety@springernature.com

Printed and bound by CPI Group (UK) Ltd, Croydon, CR0 4YY
26/03/2026
02078941-0014